The **10** Things
All Future Mathematicians and Scientists Must Know

(But are Rarely Taught)

Edward Zaccaro

Hickory Grove Press

About the Author

Ed lives outside of Dubuque, Iowa, with his wife and three children. He has been involved in education in various forms since graduating from Oberlin College in 1974. Ed has taught students of all ages and abilities, but his focus for the past ten years has been working with mathematically gifted students at the elementary and middle school level. When unable to find sufficient curriculum and materials for his students, he began to develop his own, resulting in the following collection of books.

- ♦ *Primary Grade Challenge Math*
- ♦ *Challenge Math for the Elementary and Middle School Student*
- ♦ *Real World Algebra*
- ♦ *The Ten Things All Future Mathematicians and Scientists Must Know (But are Rarely Taught)*

Ed, who holds a Masters degree in gifted education from the University of Northern Iowa, has presented at state and national conferences in the areas of mentoring and gifted education.

About the Artist

Jack Berg, who lives in Galena, Illinois, is a self-taught artist and musician. Jack started his art career working as an animator and later became the art director for a large corporation. Jack is currently the owner of Heartland Art Studios and Heartland Recording Studios.

Cover designed by Wilderness Graphics, Dubuque, Iowa.

Copyright©2003 Edward Zaccaro

Phone: 563-583-4767
E-mail: challengemath@aol.com
http://www.challengemath.com

Library of Congress Card Number: 2003103377
ISBN: 0-9679915-4-4

This book is dedicated to:

James Randi
Richard Feynman
Marilyn Vos Savant
Martin Gardner
Carl Sagan

Table of Contents

Introduction

Mathematicians and scientists have been closely tied to many famous disasters. The Challenger explosion, the failure of the Mars Explorer, and the Kansas City Hyatt Regency walkway collapse, all involved thinking errors. This book presents the ten things our future mathematicians and scientists must know to prevent these kinds of tragedies from occurring. Because science and mathematics instruction is often dominated by facts and calculation, children are rarely exposed to these important concepts. These ten concepts, and the stories that bring them to life, clearly show the strong connections between science and math and the world we live in.

Math and Science Will Tell You the Truth: Math and science are not like referees and umpires that you can argue with if you don't like what they tell you. The beauty of math and science is that they give you an objective look at a situation. It can be dangerous to ignore their message.

Your Mind can be Fooled (Counterintuitive Thinking): The solutions to many problems are counterintuitive. Your brain can easily be fooled into strongly believing in the correctness of an incorrect answer or solution.

Occam's Razor: When you are searching for an explanation for a strange or unusual event, the simplest or most down to earth explanation is usually the correct explanation.

Mistakes and Frustration are a Part of Learning and a Part of Life: Some children are more susceptible to fearing mistakes and challenges because they are foreign to them. They must understand that challenge and frustration are a part of learning and life.

It is Important to Keep an Open Mind: Many of the greatest ideas and discoveries of the last 500 years were ridiculed and attacked when they were initially presented. The medical advances that were delayed are tragic examples of the implications of a failure to keep an open mind.

It is Important to Maintain a Healthy Skepticism: It is important to maintain a healthy skepticism. You must not have a mind that is so open that your brains fall out.

Don't be Fooled by Statistics: The study of statistics is often thought of as a dry and boring endeavor. This is not true! The use of statistics has saved millions of lives. In addition, the manipulation, deception and outright lying that can accompany the use of statistics, makes it imperative that children learn how to interpret them.

You Must Know the Difference Between Cause and Correlation: The failure to understand the difference between cause and correlation has led to an inability to differentiate what is true and what is not true in medicine, education and in many areas of our everyday lives.

Ethical Decision Making: During almost all careers, individuals will likely face situations where ethics and an ability to maintain those ethics will influence the choices that are made.

Bias is Everywhere: Bias is almost everywhere. It influences the results of surveys, scientific experiments, the beliefs that we hold, and many of the decisions that we make.

Chapter 1

Math and Science Will Tell you the Truth

Math and science are not like referees and umpires that you can argue with if you don't like what they tell you. Math and science are coldly and cruelly indifferent to your hopes, dreams and wishes. They give you an honest and objective look at a situation. Do not ignore their message!!

Challenger Disaster

January 28th, 1986, was going to be a special day for NASA and the space program. For the first time a teacher was going to accompany six other astronauts aboard a space shuttle and teach lessons from space.

There was great excitement, but there also were many people who were worried. Several parts of the space shuttle had the potential to be affected by the extremely cold temperatures that were predicted for that morning. Engineers were especially concerned about the rubber-like seals in the solid rocket boosters which kept hot gases from coming out the sides.

The cold temperatures would mean that the O-rings might not be flexible enough to keep the hot gases from escaping. If the gases did escape, there could be a catastrophic explosion because the solid rocket boosters were next to a large tank which contained liquid hydrogen.

The night before the launch, NASA managers had a telephone conference with the company that made the solid rocket boosters. The engineers at the company knew about the solid rocket boosters and the O-rings, so they were asked whether it was safe to launch.

The engineers were very concerned that cold temperatures would not allow the O-rings to seal properly, which might lead to an explosion. In addition, the engineers had information from previous flights that made them even more nervous. The only other time the shuttle was launched in cold weather, there had been a significant erosion of the O-rings. That launch occurred at 53° and now they were being asked to approve a launch at a temperature of 29°!! They said no, the shuttle should not be launched.

Because the cold weather will make the O-rings so stiff that they may not seal properly, we recommend that the launch be delayed until the O-ring temperatures reach 53°F.

This meant that the launch would have to be delayed because the predicted temperature at launch was 29° F.

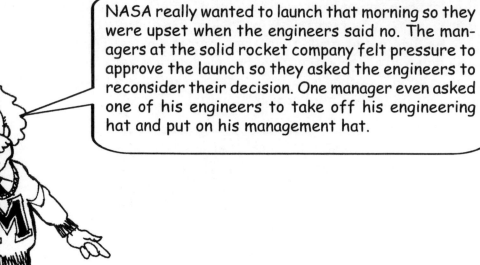

NASA really wanted to launch that morning so they were upset when the engineers said no. The managers at the solid rocket company felt pressure to approve the launch so they asked the engineers to reconsider their decision. One manager even asked one of his engineers to take off his engineering hat and put on his management hat.

I'm feeling pressure from NASA to approve this launch. I would like you to reconsider your position.

We still think it is too dangerous to launch because of the effects of cold weather on the O-rings. We still vote no!

When the engineers again recommended that the launch be postponed, their managers decided to ignore what the engineers said and told NASA that it was okay to launch. The following morning, as expected, was clear and cold. Challenger was about to be launched, even though the engineers said it was not safe to do so.

3....2....1 Challenger's powerful engines ignited and it slowly pulled away from the launch pad. Suddenly, just as the engineers feared, one of the O-ring seals failed and flames shot out the side of one of the solid rocket booster's joints. Seconds later, the hydrogen tank exploded and the seven astronauts plummeted helplessly to their deaths.

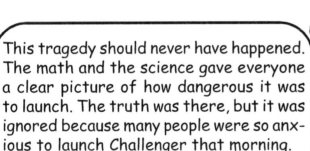

This tragedy should never have happened. The math and the science gave everyone a clear picture of how dangerous it was to launch. The truth was there, but it was ignored because many people were so anxious to launch Challenger that morning.

Math is not like a referee or an umpire that you can get mad at and accuse of favoritism, blindness or incompetence. Math is different. Math doesn't care how much you want something to be true, how many people your idea will help, or how much easier life would be if something were true. Math is coldly indifferent, but it does one thing very well. Math and science will give you a very clear view of a situation. It is up to you whether you pay attention to it or not.

Discussion Questions:

1) When the manager overruled the decision of the engineers, should the engineers have called NASA directly, or tried to talk to the astronauts directly? Why or why not?

2) How was this situation similar to the story of the mice and the mousetraps?

3) If you were an astronaut and you received a call from an engineer saying the shuttle wasn't safe to launch, would you refuse to take part in a launch that morning?

How Mathematicians Made Sure that a Trial was Fair

Juries are supposed to be picked randomly. This helps to ensure that a person charged with a crime receives a fair trial.

I am the prosecutor and I want the jury to have police officers.

I want my brothers and my mother on the jury.

It is easy to see that neither of these juries would be fair. A jury composed of relatives would almost certainly find the defendant innocent. A jury composed of all police officers would be more likely to unfairly convict the defendant.

Several African Americans who were convicted of crimes by juries that were all white, appealed their cases to a higher court claiming that their juries were not picked fairly.

The city where I live is 50% African American and 50% white. Why was my jury-pool made up of 95 whites and only 5 African Americans?

I know it looks like the jury wasn't picked fairly, but trust me, it was just a coincidence.

When these cases went to an appeals court, it had to be determined whether jury pools with 95 whites and 5 African Americans in cities that were 50% African American could possibly have happened by chance. The lawyers for the convicted men argued that these juries were not picked fairly because of racial bias. The prosecutors insisted that the juries were fairly picked. The courts had to decide who was right.

How could this question be settled? I think the only thing that can solve this problem is mathematics!

Mathman is here to save the day!

The probability that this jury pool of 95 whites and 5 African Americans was just a coincidence is lower than one in one quintillion!

This shows that it was very, very unlikely that the jury pool was selected at random. A one in one quintillion chance is a lower probability than getting a royal flush three times in a row. I think that we can safely say that there was racial bias.

Mathematics tells us that this gentleman was not given a fair trial. The defendant wins his appeal.

Discussion Questions:

1) If potential jurors for a trial were picked by someone standing outside a country club and signing up the people who walked out, would the resulting jury be fair? Why or why not?

2) In the 1930's, juries were sometimes picked by taking names from a phone book. Why would this not be a fair way to pick juries in the 1930's?

3) Today juries are usually picked from lists of registered voters. Why is this a fair way to pick juries?

Dr. Snow and Cholera

The year was 1840 and Dr. Snow was very worried. London was again hit by a terrible cholera epidemic. The disease was not only deadly, but it was mysterious as well. It would suddenly appear in cities, kill thousands, and then disappear. Dr. Snow knew that it was important to find out what caused the disease if he wanted to prevent another outbreak.

I am perplexed. Why do epidemics just appear? What causes them? Why do some people get sick while others do not?

In the year 1840, doctors didn't understand what caused diseases. They didn't know about bacteria or viruses, so they had a hard time determining why some people got sick and others didn't. No wonder Dr. Snow was confused.

Dr. Snow thought of a way to better understand the disease. He put dots on a map of London where each cholera death had occurred. After Dr. Snow finished, something became very clear to him. There was a strong connection between a water pump on Broad Street and those who died of cholera. When he looked further at the map, he found something that strengthened his theory even more. Dr. Snow found one cholera death that was far away from the Broad Street pump. When he talked to relatives of the woman who died, he discovered that she liked Broad Street water so much that she paid someone to bring the water to her home.

Dr. Snow didn't know why or how, but he knew that something was contaminating the water pump on Broad Street and killing thousands of people. After considering his options, Dr. Snow broke the handle of the pump on Broad Street. Just as he predicted, the cholera epidemic ended when people stopped getting water from that pump.

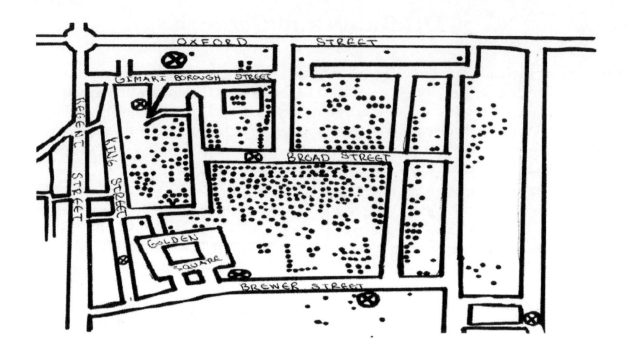

 Waterpump

I used math to help save the lives of many people.

The statistics Dr. Snow gathered about cholera deaths in London not only taught us the importance of sanitation in preventing illness, but it also showed us the importance of organizing and analyzing data. This story makes it very clear that Dr. Snow's use of statistics led to changes that saved thousands of lives.

Many people think the study of statistics is boring. This story not only shows how interesting statistics can be, but it also shows the important uses of statistics.

If you want to learn how people lie with statistics, read chapter 7.

Discussion Questions:

1) Dr. Snow broke the handle on the Broad Street pump when his research showed that its water was probably causing cholera. Did he do the right thing by acting on his own, or should he have explained his theory to the leaders in his community and tried to convince them that the pump should be disabled?

2) In 1840, people didn't know what caused diseases because they did not know about bacteria and viruses. How do you think they explained diseases in 1840?

3) Why do you think the Broad Street pump was contaminated and the other pumps were okay?

The Tragic Death of Aaliyah

Aaliyah was a popular singer who many people thought was well on her way to stardom. In August of 2001, Aaliyah was in the Bahamas to film her latest music video. When filming was completed, Aaliyah and eight others boarded a small plane for a short flight to Miami, Florida. Shortly after takeoff, the plane banked sharply and then crashed, killing all nine people aboard.

When the investigation began, it immediately became clear that there was more to this crash than mechanical failure. It was reported that before takeoff, the pilot asked the passengers to leave some of their luggage behind because the plane was dangerously overloaded. The passengers refused to remove any luggage, so the pilot gave in and tried to fly the overloaded plane.

There is too much weight on the plane. There are too many people and too much luggage.

We don't want to rent another plane and we need to take all our luggage.

The small plane was a Cessna 402B, which was certified to carry six to eight people and had a maximum allowable take-off weight of 6300 pounds. Because the plane and its seats weighed 4100 pounds, the luggage, fuel, and passengers had to total less than 2200 pounds.

> The fuel weighed 800 pounds and the luggage weighed 575 pounds. That meant that the weight of the nine passengers had to be less than 825 pounds.

> We know that Aaliyah's bodyguard weighed close to 300 pounds, so the other eight people on the plane would have had to weigh less than 525 pounds. That would mean that each adult's weight would have had to average approximately 65 pounds.

When the weight of the plane, fuel, luggage, and passengers was totaled, it became clear that the plane was overloaded by at least 700 pounds. Due to the excess weight, the plane could not stay airborne and crashed a short distance from the airport.

Once again, math and science were ignored because people did not like what the math and science told them.

The result was that nine people died needlessly.

Discussion Questions:

1) In one sentence, write what the pilot should have said to the passengers when they refused to listen to his warning about the plane being overloaded?

2) Why do you think the passengers ignored the warning and insisted on keeping everyone and their luggage aboard?

3) Why do you think the pilot gave in and tried to fly the overloaded plane?

The Truths of the Universe are Revealed Through Mathematics

Almost 2500 years ago, mathematics changed the study of space from one of fantasy and guesswork to a real science. No longer did we need to guess whether the Earth was flat or round, or whether the sun was smaller or larger than the Earth. Mathematics gave us answers to those questions. Mathematics even revealed the amazing truth about the great distances in space.

As the mathematicians of that time sought to use mathematics in their study of the universe, one crucial piece of information eluded them. They needed to determine the size of the Earth. They knew that once they had that information, they could unlock many of the mysteries of the universe.

I wonder if the moon is a mile away, a 100,000 miles away, or a million miles away?

Eratosthenes, who lived in Greece over 2500 years ago, used mathematics in a clever way to not only prove that the Earth was round, but also to finally determine the size of the Earth.

It is a myth that people thought the Earth was flat in the time of Columbus. Educated people had known for hundreds of years that the Earth was round.

Wait until you hear how Eratosthenes found out how big the Earth was. It was so clever that it gives me the chills.

Eratosthenes knew that the distance between the cities of Alexandria and Syene was 500 miles. He also knew that when the sun was directly overhead in Syene (during the summer solstice), it was 7 1/2° away from being directly overhead in Alexandria.

Eratosthenes also knew one other thing that made it possible to determine the circumference of the Earth. He knew that there were 360 degrees in a circle and that this 500 miles was a small 7 1/2° slice of a big Earth pie. Because 360 ÷ 7 1/2 is equal to 48, Eratosthenes knew there were 48 slices of this 7 1/2° size pie in the Earth.

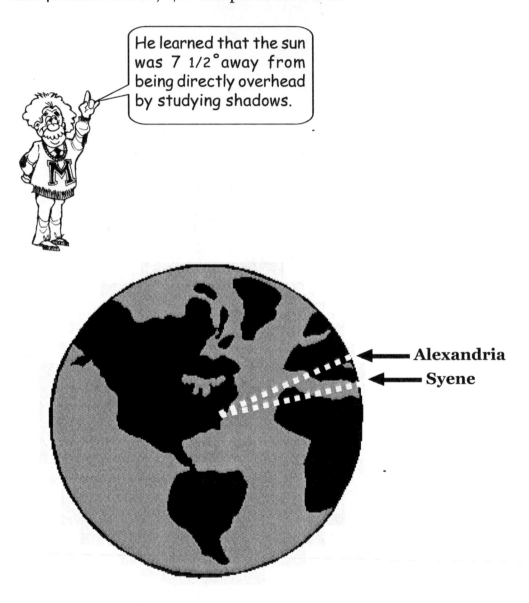

From this information, it was an easy calculation to show that the circumference of the Earth was approximately 24,000 miles. Once the circumference was determined, an estimate of the diameter of the Earth was another easy calculation: 24,000 ÷ π = 7643. The Earth's diameter was about 8000 miles.

Several years after this brilliant discovery, a Greek mathematician named Hipparchus used trigonometry to determine the approximate distances to the sun and the moon. He came very close to the moon's actual distance, but he underestimated the sun's distance by 80,000,000 miles.

Before Eratosthenes's discovery, most people thought the moon and the sun were a short distance from the Earth. Also, because they looked the same size, many people thought that the sun and moon were the same size.

Mathematics sure tells us a lot about the universe we live in!

Discussion Questions:

1) If the sun was 20° away from being directly overhead, what would Eratosthenes have calculated as the circumference of the Earth?

2) How did Eratosthenes use the fact that the sun was 7 1/2 ° away from being directly overhead in Alexandria to strengthen his theory that the Earth was round?

3) What is the summer solstice?

Math and Science Will Tell you the Truth
Level 1

1) Which telescope is a better bargain, one that cost $212 in a state with a 6% sales tax, or one that cost $198 in a state with a 12% sales tax?

2) A hot-air balloon can safely hold 1055 pounds. It currently has 6 people in it whose average weight is 128 pounds. In addition, it has a 4-foot by 6-foot metal floor that weighs 8 pounds per square foot. How many 25-pound bags of sand can be safely placed in the balloon?

3) Stan needed an average score of 70% to pass a Spanish class. He received a score of 65% on his first test and a 67% on his second test. What must Stan get on his third test to pass the class?

4) A check cashing company will cash your check for a 12% fee. What would be the charge for a $50 check?

5) A family has four children who are all boys. They want to know the probability of their next child being a girl. What is their probability of having a girl if they already have four boys?

Math and Science Will Tell you the Truth
Level 2

1) Matt needs to bring a roll of carpet home on the top of his car. The weight limit for the roof rack is 150 pounds and the carpet weighs .65 pounds per square foot. If the carpet is 6 yards long and 5 yards wide, is it safe to put the carpet on the roof?

2) A very tall tree is going to be cut down. The people who are cutting down the tree must determine how tall it is because there is a house 92 feet away that they are afraid of damaging. They measured the length of a yardstick's shadow and found that it was 28 inches long. Next they measured the tree's shadow and found that it was 72 feet. Can they safely cut down the tree?

3) If a cubic foot of snow weighs 5 pounds, how much does a cubic yard of snow weigh?

4) A 13 millimeter crack (due to metal fatigue) has been spotted in a metal beam that is an important part of the support structure for a bridge. Engineers determined that when the crack reaches 1411 millimeters, the bridge will be in danger of collapsing. The engineers took measurements each day and wrote them in a note book. If the crack continues to grow at this pace, when will the crack reach 1411 millimeters?

Day 1 - 13 mm
Day 2 - 19 mm
Day 3 - 25 mm
Day 4 - 31 mm
Day 5 - 37 mm

5) A 12,000 pound truck is being filled with sand. The container on the back of the truck is 16 feet long, 10 feet wide and 8 feet deep. The driver wants to take as much sand as she can, but there is a bridge on her route that has a weight limit of 28,000 pounds. If the sand weighs 25 pounds per cubic foot, how high can the driver fill the container on the back of the truck?

Math and Science Will Tell you the Truth
Einstein Level

1) Lindsey is trying to convince her friend to stop buying lottery tickets. Lindsey's friend is thinking about spending most of her savings account on lottery tickets because a winning ticket is worth $300,000,000. In this lottery, a player picks six numbers from 1 to 48. If all six numbers are winning numbers (in any order), the player wins $300,000,000. Help Lindsey convince her friend that her chances of winning are so small that it isn't worth spending her money on lottery tickets. What are the chances of winning this lottery if you buy one ticket? (Hint: The first number a player selects has a 6 in 48 chance of being one of the winning numbers.)

2) Mark wants to buy a sports car, but he is very concerned about the mileage the car gets. The sports car gets 12 miles per gallon, while a more practical car gets 28 miles to the gallon. If Mark is planning to drive his car 80,000 miles, how much more expensive will gas be for the sports car than for the more practical car if gas cost $1.50 per gallon? (Round to the nearest $10.)

3) A royal flush is a hand in cards that is very rare. It consists of the 10-jack- queen-king-ace of the same suit. What is the probability of getting a royal flush if you are dealt 5 cards?

4) Nathan charged $400 on a credit card that charges 22.9% interest. Nathan didn't make any payments to the credit card company for 10 years. Now he wants to pay what he owes. Assuming there are no late fees, how much money does Nathan owe the credit card company now? (Remember that Nathan will owe interest on interest each year.) Round to the nearest dollar.

5) To estimate how fast money in a savings account will grow, simply divide the interest rate into 72. The resulting number tells you how many years it takes for money to double. If a baby is given $1000 at birth, and the money is placed in a savings account that pays 12% interest, how much money will the child have when she is 66 years old?

Chapter 2

Your Mind can be Fooled

(Counterintuitive Thinking)

The solutions to many problems are counterintuitive. Your brain can easily be fooled into strongly believing in the correctness of an incorrect answer or solution.

The Kansas City Hyatt Regency Hotel Disaster

On July 17, 1981, 2000 people gathered in the lobby of the Kansas City Hyatt Regency Hotel to either watch or take part in a popular dance contest. In addition to those in the lobby, there were hundreds of people dancing on the walkways suspended high above the lobby. No one knew that they were just minutes away from one of the most serious structural failures this country had ever experienced.

As the dancers moved to the big band sound, a loud crack was heard. A split second later, two of the crowded suspended walkways collapsed onto the people below. As people hurried to get out of the way, they were showered with tons of concrete and metal. Despite the heroic efforts of rescue personnel, 114 people died and over 200 were injured. When the investigation began, it quickly became apparent that there was a major flaw in the design of the walkways.

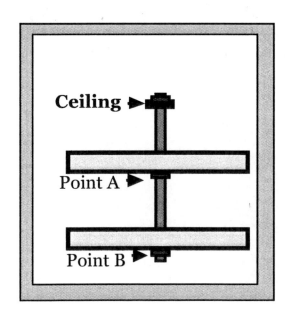

The original design called for two of the walkways to be hung with a 50-foot rod extending from the ceiling and through both walkways. In this design, the weight of both walkways was carried to the ceiling because points A and B each held up one walkway.

You can think of this design as similar to three people climbing a rope in gym class. Each person holds himself on the rope, and all the weight goes to the ceiling.

Changed Design
Ceiling →
Point A →
Point B →

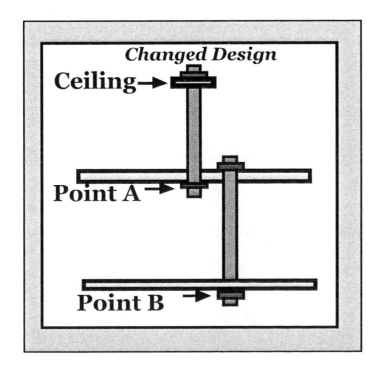

Unfortunately, somebody saw the 50-foot rod in the design for the walkway and decided that two 25-foot rods would be much easier to work with. And of course, everyone knows that two 25-foot rods are the same as one 50-foot rod. At first glance it might seem that way, but it is not the same in this situation. No longer is all the weight going to the ceiling. Now point A not only has to hold up its own walkway, but you can see from the diagram that point A is also holding up the walkway that is below it. Point A was never designed to hold the weight of two walkways. The changed design was much easier to assemble, but it led to the catastrophic collapse.

You can think of this design as similar to three people climbing a rope in gym class. But instead of each person holding the rope, each person holds on to the legs of the person that is higher up. In this situation, all the weight goes to the ceiling, but the top person's hand is holding the weight of three people.

Because point A was forced to support the weight of two concrete walkways and the additional weight of scores of people, it failed.

The human mind is such a powerful problem solver that its susceptibility to being fooled is often overlooked. If a 50-foot rod is a problem, we can use two 25-foot rods. It is easy looking back to see that this is a mistake, but at the time, it was overlooked by the architectural firm and the engineers.

Discussion Questions:

1) Can you think of a way to have the two 25-foot rods, but still have the same strength as one 50-foot rod?

2) If a construction worker noticed that the design change was dangerous and suggested that they keep the original 50-foot rod, how do you think the architect would have reacted? Would the worker have been taken seriously or just ignored?

3) The firm that designed the walkway was obviously sued for a substantial amount of money. Do you think, in addition to paying money to settle the lawsuits, that the architect who designed the walkway should have his architectural license taken away permanently?

The Game Show Dilemma

A well known game show had contestants try to win valuable prizes by picking one of three doors. The game show host knew where the valuable prize was, so after a contestant made her pick, the host would open a door he knew didn't contain the prize and show the contestant what was behind that door. He would then ask the contestant if she wanted to trade her door for the remaining door.

You picked door number 2. Let's open door number 3 and show you what you missed. Do you want to keep door 2 or switch to door number 1??

Oh, I don't know. I think I'll stay with my first pick because I have a strong feeling that door number 2 is where the prize is.

At first glance, it appears that it doesn't matter whether the contestant trades doors or not. Most people would say that each door now has a 1 in 2 chance of winning. That is not correct!

This problem is so counterintuitive that some of the smartest people in the world not only answer incorrectly that the contestant's chances are 1 in 2 for both doors, but they refuse to believe the correct answer when they hear it! The correct answer is that if the contestant keeps her original door, her chances of winning are 1 in 3. If she switches doors, her chances of winning would be 2 in 3.

> I have two degrees in math, three in science, and so much confidence in my intellectual ability that I cannot see that I am wrong.

The correct answer is easier to see if you pretend that there are a thousand doors and the valuable prize is behind one of the doors. You are the contestant and pick door number 612. Your chances of winning are of course 1 in 1000. Now the game show host, who knows where the prize is, opens the rest of the doors except door number 179. He now offers you the chance to change doors to number 179. Should you switch? You would be wise to switch because door number 179 has a 999 out of a thousand chance of winning, while your door still has only a 1 in 1000 chance of winning.

> Please keep me, I'm your lucky door!

> I'm not a lucky door, but I have a 999 out of 1000 chance of winning.

If you still don't get it, you are in good company. Most people have a hard time believing the right answer. The part that is very important in this problem is that the game show host knows where the prize is. If he didn't, then the obvious answer would be the correct answer.

Unfortunately, hundreds of contestants on this game show did not know the correct way to determine the probability for each door. If they had, they would have switched every time and doubled their chances of winning the big prize.

Discussion Questions:

1) If a contestant has a very strong feeling that her original choice of doors is correct, should she take that into account when she tries to decide whether to switch to a different door?

2) Why would the probability be different if the game show host did not know where the prize was?

3) Why do you think this problem is hard to answer correctly? Why is the answer hard to see even when it is given to you?

The Strange World of Average Speed

Mark rode his new bike at a speed of two miles per hour on his way to school. Because Mark's school was only a mile away from his house, the trip took him 30 minutes. Mark's friends thought it was funny that Mark rode his bike at that speed, so they teased him the entire day about how slow his bike was.

> I could have gone much faster if I wanted to.

During science class, Mark was defending his bike so much that his friend Venus couldn't take it anymore. She challenged Mark with a simple bet. "Your bike is so slow that I don't think you can ride it fast enough on your way home to average 4 miles per hour for your round trip to school and back. If you win, I'll buy you any bike you chose. If you lose, then you have to buy me any bike I chose."

Mark knew about averages, so he felt pretty confident that he would win the bet. He knew that when he scored 80% and 60% on two tests, their average was 70%. Mark figured that all he had to do was ride his bike at a speed of 6 miles per hour on his way home to average 4 miles per hour.

> I'm glad you're confident Mark. I'll see you in class tomorrow. In case you're wondering, the bike I want is on sale for $800. I hope you have a good allowance Mark.

> I'll take that bet! I'll have you know that I can peddle this bike at a speed of 30 miles an hour when I want to. Start saving your allowance Venus because the bike I want is very expensive!

Unfortunately for Mark, he could never average 4 miles per hour on his round trip between home and school. Average speeds are not figured the same way as test scores are.

I beg to differ, Einstein. They really are figured the same way, but it is a little more complicated. Because you spend more time traveling when your speed is slower, you have to count the slow speed more.

To find the average speed, you use the formula **Distance = Speed x Time**. In this case, we know the distance is 2 miles and the average speed we want is 4 miles per hour. We can plug those numbers into the formula. We'll use *t* for time because we are trying to find out how much time it takes to average 4 miles per hour.

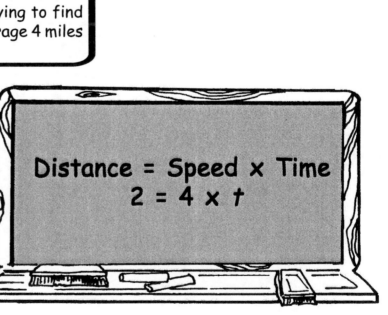

Distance = Speed x Time
$$2 = 4 \times t$$

When the equation is solved, it is clear that *t* must equal 1/2 hour. Because Mark has already taken up that 30 minutes on his way to school, he can never average 4 miles per hour -- even if he rode his bike at a speed of 1000 miles per hour.

Discussion Questions:

1) A hiker goes up a mountain path at an average speed of 3 miles per hour. If she then goes down the same path at a speed of 6 miles per hour, what was her average speed for the trip?

2) If speed averages were found the same way that test scores are determined, then someone who walks 1 mile to the airport at a speed of 2 miles per hour and then travels 3000 miles on a jet at a speed of 500 miles per hour, would have an average speed of 251 miles per hour for the trip. Why would an average speed of 251 miles per hour not make sense for this trip?

3) A canoeist took a trip downstream at a speed of 6 miles per hour. He then returned to the starting point at a speed of 2 miles per hour. What was his average speed for the trip?

The Stranded Astronaut

Suppose you are an astronaut who is on a space walk. A 100-foot tether is keeping you attached to the space shuttle and the only thing you are carrying is a battery-operated fan. Suddenly, a meteor slices through your tether and you are left stranded 100 feet from the safety of the shuttle. How would you get back to the shuttle?

If you are thinking that you can turn on the fan and have it blow in the opposite direction, you would be unsuccessful at returning to the shuttle. If you are thinking that a swimming motion will help return you to the shuttle, that idea will also fail.

If you thought of these ways to return to the shuttle, your mind is being influenced by experiences you have had on Earth. You must think in a counterintuitive manner to save yourself.

Because there is no air in space, the fan and your moving arms would not have anything to push against.

Newton's laws of motion can help you devise a way to return to the space shuttle. Newton's third law says that for every action there is an equal and opposite reaction. If a shotgun is fired while it is against someone's shoulder, the gun will fly back into the shoulder as the bullet is propelled forward. Using this law, you could throw the fan in the opposite direction that you wanted to go. This would move you back toward the shuttle.

What if the fan is not thrown hard enough? Will the astronaut go only halfway back and be stuck with nothing left to throw?

The astronaut could use Newton's first law to ease his mind. This law states that an object in motion tends to stay in motion at constant speed in a straight line. If throwing the fan sends you toward the shuttle at 10 mph, you will keep moving in that direction and at that speed until you run into the shuttle.

If you miss the shuttle though, you will go in that direction and at that speed until you run into the gravitational field of an object in space.

Discussion Questions:

1) Why doesn't a ball travel forever when it is thrown on Earth?

2) In a scene from a movie, a dog that dies is thrown out of a spaceship. Ten minutes later it is traveling right next to the spaceship. Why does this violate one of Newton's laws of motion?

3) Would a parachute work on the moon? Why or why not?

Your Mind can be Fooled
Level 1

1) A ball and a glove together cost $10. If the glove cost $9 more than the ball, what is the cost of the ball?

2) Some scientists estimate that the universe contains 10^{87} atoms. How many atoms are there in one tenth of the universe?

3) Which causes more pressure on a rope?

    a) Ten 200-pound people on one side of a rope who are pulling with all their might with the other side attached to a tree.

    b) Ten 200-pound people on one side of a rope with another ten 200-pound people on the other side. Both sides are pulling with all their might.

4) Why did Galileo have so much difficulty convincing people that the Earth revolved around the sun?

5) You are at the top of a very tall building with a marble and a bowling ball. If you drop them, which one will hit the ground first? Why?

Your Mind can be Fooled
Level 2

1) Which collision would be more violent?

a) A car traveling 60 miles per hour hitting a brick wall.

b) Two identical cars traveling in opposite directions at 60 miles per hour and hitting each other head on.

2) Phil hiked up a mountain at 3 miles per hour and then went down the same trail at a speed of 6 miles per hour. What was Phil's average speed for his hike?

3) A car went around a one mile track at a speed of 30 miles per hour. At what speed must the car travel on its second time around the track if the driver wants to average 60 miles per hour for the two mile trip?

4) Mark started a new job at a salary of $20,000 per year. He was given a choice as to how he would receive raises in salary during his time working with the company. What option should Mark choose to receive the most amount of money?

Choice 1: A raise of $1000 each year
Choice 2: A raise of $400 each six months

5) Chris, Laura, and Stacie were three friends who went to college and medical school in order to fulfill their dreams of becoming doctors. As they were finishing their work and getting close to graduation, the dean of the medical school put their names in a hat, picked one, and announced that the person whose name he picked would be expelled from school. (He did this because he was mean and because he liked probability problems.)

Chris was so upset that he begged the dean to look at the name he picked and tell him the name of one his two friends who would be allowed to graduate. The dean said that Laura would be graduating. Now Chris was even more upset. "The probability that I will be kicked out of school is now 1 in 2 instead of 1 in 3." Was Chris correct about his probability of being expelled from school? Why or why not?

Your Mind can be Fooled
Einstein Level

1) Three coins are in a hat. One of the coins has heads on both sides, while another coin has tails on both sides. The third coin has heads on one side and tails on the other. You pick a coin and look at one side and see that it is heads. What is the probability that the other side will be heads?

2) Two basketball players are each 9 for 30 in shooting baskets. Player A then goes 1 for 10 on his next 10 shots. Player B does much better and sinks 6 of his next 30 shots. Even though Player B did much better, both players ended up shooting 25%.

Player A: 10 for 40
Player B: 15 for 60

How can it be that Player B has the same percentage as Player A?

3) You see two boxes on a table. You are told that one box has twice as much money as the other box. You open the box of your choice and find that it contains $900. Now you are given the opportunity to switch boxes. Do the laws of probability suggest that you switch?

4) You meet a long lost friend and find out that she has two children. You ask her the following question:

"Is at least one of your children a boy?"

Your friend responds that yes, at least one of her children is a boy. What are the chances that her other child is also a boy?

5) Twenty-three children are in a classroom. On the first day of school, the teacher asks what the chances are that two children in that class share the same birthday.

a) Very unlikely
b) Small chance
c) About a 50-50 chance
d) Almost certain

Chapter 3

Occam's Razor

If you are searching for an explanation for an event that is perplexing or unusual, the correct explanation is usually the most basic or down to earth explanation. - Occam's Razor -

Remember that Occam's Razor tells us to think of the simplest explanation first. It doesn't say that the simplest explanation is always the right explanation.

I'm glad you said that because occasionally there is a bizarre explanation that really is the correct explanation.

I remember one time I thought that I was abducted by aliens, but I used Occam's Razor and decided that the most likely explanation was that I had a nightmare.

Clever Hans the Horse Genius

Wilhelm Von-Osten was a Russian aristocrat who lived in Germany. He had an interesting theory that concerned the intellectual abilities of animals. Von-Osten thought that the only reason animals were not as smart as humans was because they didn't go to school.

Wilhelm Von-Osten, who took great pride in his teaching abilities, started math classes for his horse. Soon, Hans was using his hooves to tap out answers to addition, subtraction and multiplication problems. Month after month, Hans would tap out correct answers and Von-Osten would reward him with praise and bites from a carrot. It appeared that Clever Hans had become a horse genius.

Word of the equine genius spread like wildfire. Soon thousands of people came to see Clever Hans solve math problems. Over and over again, Clever Hans would tap out the correct answers to the problems that were given to him. Even German scientists, who were initially skeptical, were starting to think that maybe a horse could be taught math.

Not everyone was convinced though. A scientist named Oskar Pfungst decided to use the scientific method to see if Clever Hans really was able to solve math problems. The first part of the test was to see if Clever Hans could solve problems when his owner wasn't in the room. Surprisingly, Clever Hans did almost as well when Wilhelm Von-Osten was not in the room. Another piece of information that Pfungst wanted to know was if the horse could answer questions when no one in the room knew the question.

Pfungst had Wilhelm Von-Osten hold up cards with numbers on them. Clever Hans easily tapped out the numbers with his hoof. Next, Wilhelm Von-Osten was instructed to hold up cards without looking at the numbers that were on the cards. When Von-Osten did not know what number was on the card, Clever Hans was unable to tap out the number. Clever Hans was not the horse genius that everyone thought he was.

Even though Clever Hans could not really solve math problems, he still was very clever. It appears that the people watching Clever Hans tap out answers were somewhat tense because they didn't know whether he would answer the question correctly. When he tapped his hoof the correct number of times, the spectators would relax slightly. Clever Hans had learned that if he stopped tapping when this happened, people would cheer and he would get carrots.

I may not be able to really solve math problems, but I am clever. Can I have a carrot?

Don't feel bad Clever Hans. People are doubting my abilities all the time because I'm a rodent.

Occam's Razor would warn us that there was probably something fishy going on if a horse appeared to be able to solve math problems.

Discussion Questions:

1) Do you think that Wilhelm Von-Osten accepted the results of Pfungst's scientific study?

2) Pretend you called a psychic hotline and the psychic told you something that was true. Use Occam's Razor to explain how she knew this piece of information.

3) What other experiments could have been done to determine whether Clever Hans really was solving math problems?

The Cold Fusion Fiasco

The Earth is warm because of the high-temperature fusion of hydrogen nuclei that is taking place deep within the sun almost 93,000,000 miles away. For many years it has been a dream of scientists to somehow devise a way to duplicate this kind of fusion on Earth. If this could be done, we would have an almost limitless supply of pollution-free energy.

Nuclear fusion is different from the nuclear reaction that takes place at nuclear power plants. That type is called nuclear fission and is much easier to accomplish on Earth. In a fusion reaction, the reacting nuclei both have positive charges. The natural repulsion between them must be overcome before they can join.........................

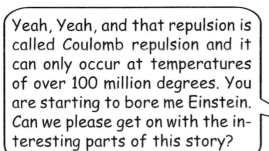

Yeah, Yeah, and that repulsion is called Coulomb repulsion and it can only occur at temperatures of over 100 million degrees. You are starting to bore me Einstein. Can we please get on with the interesting parts of this story?

On March 23, 1989, a dramatic announcement came from the University of Utah. Two chemists, Martin Fleischman and Stanley Pons, had discovered a way to efficiently produce nuclear fusion. They called their technique cold fusion because it didn't involve the high temperatures that conventional attempts at nuclear fusion had depended on.

If their discovery was true, it would not only be the discovery of the century, but a case could be made that it would be the greatest scientific discovery of all time. What made this claim so hard to ignore were the credentials of the scientists who were making the claim. Fleischman, who was a member of the British Royal Society, was a visiting professor from England. Pons, a published researcher, was a respected professor of chemistry at the University of Utah.

Well-known astronomer Carl Sagan, once said that extraordinary claims require extraordinary proofs. If ever a claim was extraordinary, this one was!

Occam's Razor warns you to rule out simple explanations before you make extraordinary claims. I sure hope the two professors eliminated all the simple explanations for the appearance of nuclear fusion. If they didn't, they are going to look awfully foolish.

After weeks of news coverage and debate among scientists from around the world, the truth finally came out. Unfortunately for Fleischman and Pons, simple explanations for the appearance of nuclear fusion were not ruled out before they made their dramatic announcement to the world. When their claims were carefully studied, it was found that cold fusion not only violated several accepted scientific principles, but it also became evident that what appeared to be nuclear fusion was easily shown to be the result of elementary flaws in the experiments that the two men conducted.

Even though cold fusion was discounted by other scientists, Fleischman and Pons had a very hard time giving up on their discovery. They issued a statement that said that critical scientists will end up "eating a lot of crow."

Remember, Occam's Razor suggests that when you are searching for an explanation for something that is strange or unusual, simple explanations should be ruled out before extraordinary claims are made.

How embarrassing for the scientists. I wonder why they didn't use Occam's Razor.

If you are searching for an explanation for an event that is perplexing or unusual, the correct explanation is usually the most basic or down to earth explanation. **- Occam's Razor -**

Discussion Questions:

1) Why was it very hard for the two scientists to admit that they were wrong?

2) Why do you think that Fleischman and Pons didn't carefully check their work before they announced their discovery to the world?

3) Pretend you are Fleischman and Pons. Write two sentences that will be given to the press admitting your mistakes and apologizing for not only making the mistakes, but also for taking so long to admit them.

Bigfoot

Beginning in the 1800's, sightings of apelike creatures were being reported in many parts of the world. But it wasn't until 1958, when a bulldozer operator found large footprints circling his machinery, that the Bigfoot legend took hold in this country.

Americans, who for years had heard stories from other countries about creatures known as Yeti, Sasquatch, and the Abominable Snowman, took quickly to this new creature. Over the next 40 years, numerous large footprints were found in different parts of the country. In addition, there were many eyewitness sightings of Bigfoot-like creatures and even films that appeared to show Bigfoot in his natural setting.

As we try to determine whether Bigfoot is real, we need to keep Occam's Razor in mind. The true explanation for each footprint, sighting, or film image is probably the most simple explanation.

I would also say that we can use Carl Sagan's rule again. "Extraordinary claims require extraordinary proofs."

The simple explanation for the many Bigfoot footprints is that they are hoaxes. Some prints have had two toes, while others had five. Some have had toes that were pointed forward, while others were pointed at angles. If there was a new species of apelike creatures, they would undoubtedly leave footprints that were fairly consistent as to the number of toes and the direction the toes pointed.

So it appears that the footprint evidence is weak.

The eyewitness sightings of Bigfoot are probably the most unreliable evidence that is available. Even if the people who reported seeing Bigfoot-like creatures were not involved in a hoax, it is well-known to psychologists that people are usually not good at accurately reporting what they see. Many court cases that were dependent on eyewitness testimony have fallen apart under cross-examination.

In addition to inaccurate identifications at crime scenes, people are notorious for mistaking known animals and objects for monsters and alien vehicles.

Films of Bigfoot-like creatures have been responsible for convincing millions of people that Bigfoot really does exists. Occam's Razor is a very important tool when one tries to determine whether these films show a new species of animal or are simply hoaxes. One important question to ask is whether it is possible that the "creatures" in the films are really men in ape-suits. In one of the most famous Bigfoot films, experts say that because of the size and gait of the creature, it is almost certain that it is a person in an ape-suit. One expert even claims that he has spotted a bell-shaped costume fastener on the hip of the Bigfoot in the film. If it is possible that a man is in that suit, then Occam's Razor would tell you that it is very likely that a man **is** in the suit.

One thing that makes me very suspicious is that the person who took the most famous Bigfoot film made a lot of money from it.

After close to 50 years of debate over the existence of "Bigfoot", the truth finally came out concerning the original large footprints that circled the bulldozer in 1958. After the death of Ray Wallace in late 2002, his family admitted that he was responsible for the footprints that launched the Bigfoot craze. They reported that a friend carved the feet for Mr. Wallace, and a few days later, he went with his brother and made the tracks near the machinery.

It was reported that he made the footprints as a prank, but after so many people believed that they were from a Bigfoot, he was afraid to admit his part in the joke because he was afraid that his friends would get mad at him.

> Sorry about that, but I had no idea how gullible people would be.

The disclosure of Ray Wallace's prank has not deterred those who believe in Bigfoot. Because legends usually endure, it is highly unlikely that believers in Bigfoot will ever be deterred--- no matter what the evidence. When the question of Bigfoot is looked at in a rational manner though, the only reasonable conclusion that can be drawn is that it is very unlikely that Bigfoot does exist. If there really were undiscovered creatures roaming the United States, there's a good chance that one would be hit by a car, shot by a hunter, or would die naturally and be discovered. As yet, there has never been a body or skeleton found that would prove the existence of Bigfoot-like creatures.

> Remember that when you are judging a claim such as those made about Bigfoot, it is the quality of the evidence that is important, not the quantity.

Discussion Questions:

1) What does it mean when Einstein says that it is the quality of the evidence and not the quantity that is important?

2) Why do you think that so many people believe in Bigfoot?

3) Should pranks like Ray Wallace's be against the law?

Crop Circles

There have been hundreds of reports of crop circles in the last 15 years. Many people believe that some have been made by extraterrestrials visitors, while others think that the crop circles are nothing more than elaborate pranks. As we seek to find the truth behind this mystery, it is important to keep Occam's Razor in mind.

Occam's Razor suggests that we look at the more down to earth explanations before we consider extra-terrestrial visitors.

One of the reasons that millions of people believe that crop circles are more than pranks is because of eyewitness reports of strange occurrences that have accompanied the discovery of the crop circles. Gracefully swooping flying saucers, hovering space vehicles, or beams of colored lights, are some of the activities that have been reported by witnesses. These eyewitness accounts appear at first glance to be fairly strong proof of the existence of visitors from outer space, but it is very likely that the eyewitnesses were misled in some way or another. Although they dutifully reported what they saw, there are many simple explanations for even the most bizarre accounts.

I made a list of several things that have been mistaken for alien visitors.

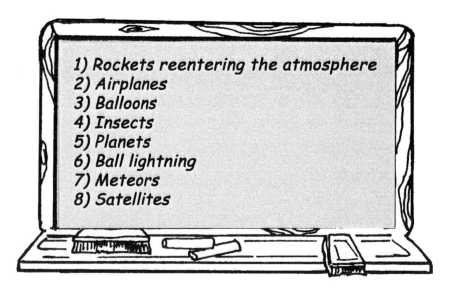

1) Rockets reentering the atmosphere
2) Airplanes
3) Balloons
4) Insects
5) Planets
6) Ball lightning
7) Meteors
8) Satellites

There are two reasons why people are easily misled by strange occurrences. One, things that are unfamiliar can easily be seen as unnatural. Two, the more you want something to appear, the easier it will be to see it.

The first report of crop circles happened in the middle 1970's along the coast of Southern England. Many people insisted that it was impossible for some of the patterns to be made by humans, while others said that they were clearly hoaxes perpetuated by an individual or individuals who very much enjoyed the attention that their activities brought.

As England debated the origins of the strange patterns, more and more pictograms were found in wheat, oat, or barley fields. Soon the phenomenon spread to other countries. Books were written on the subject, magazine articles appeared, and scientific research was undertaken as interested parties debated the origins of the strange patterns.

This escalation of reports would tend to reinforce the view that these patterns were a series of hoaxes. Publicity tends to encourage copycats.

In 1991, two men from Southhampton, England, named Doug Bower and Dave Chorley, publicly announced that they were the ones who started the crop-circle craze. They claimed to have made the initial patterns in the mid-70's and then continued with the activity for 15 years. The two concocted the idea of crop circles in an attempt to fool the UFO-crazed population. It is now clear that they succeeded beyond their wildest dreams.

Their method was a simple one that involved the use of a rope and a long plank of wood. With these tools, they carved the immense circles in local wheat fields that mesmerized the country. As time went on, Bower and Chorley made increasingly more complex designs.

Sorry to fool everyone, but you were so easy to fool. You shouldn't have been so credulous.

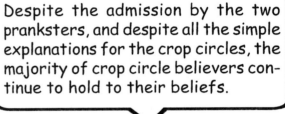

Despite the admission by the two pranksters, and despite all the simple explanations for the crop circles, the majority of crop circle believers continue to hold to their beliefs.

It is interesting to note that in the beginning, no one noticed their late-night shenanigans and they almost abandoned their crop-circle hoax. But when the public and the press took an interest, Bower and Chorley were encouraged to continue the activity.

Discussion Questions:

1) Do you agree with the statement that publicity encourages copycats? If you do, do you think some things should not be reported because of the chance that others will copy the behavior?

2) Do you think that Bowers and Chorley's prank was harmless?

3) Some people criticize the use of Occam's Razor because they say it rules out new discoveries. Explain how the proper use of Occam's Razor does not rule out new discoveries.

Occam's Razor
Level 1

1) Occam's Razor says that if you see hoofprints, think horses not zebras. What does this mean?

2) At one time, P.T. Barnum had an exhibit of a "Woolly Horse". He said that it was discovered in a valley deep in the Rocky Mountains. Was it really a "Woolly Horse?

3) P.T. Barnum displayed a so-called rare white elephant from Siam. He called it the "Light of Asia". It was a very popular exhibit until an unexpected storm caught the "Light of Asia" outside during torrential rains. Using Occam's Razor, suggest an explanation for this white elephant.

4) P.T. Barnum had a museum in New York City that had large signs that said : "This way to the Egress." "Follow the signs to the amazing Egress". When patrons finally went through the door to the amazing Egress, they were usually upset. Why?

5) A organization claims that it can teach you to levitate. (When a person levitates, it means he hovers above the ground.) As proof, they have a picture of someone in a sitting position who appears to be floating 6 inches above the ground. The picture is not doctored in any way. Use Occam's Razor to make a guess as to how such a photograph could have been taken.

Occam's Razor
Level 2

1) What does the word credulous mean? Give an example of someone being credulous.

2) On October 30th, 1938, people who were listening to music on the radio heard a news bulletin about an invasion from Mars. It was reported that city after city fell to the Martian invaders. It didn't take long before thousands of people on the East Coast were terrified. Newspapers sent reporters and photographers to cover the story as anxious relatives frantically called each other to see if they were okay. As it turned out, there was no invasion from Mars. What happened to frighten so many people on that October day?

3) In the 1890's, members of the British Society for Psychical Research were doing research on whether ESP really existed. One day, two young gentlemen named Blackburn and Smith, presented themselves to the society and claimed that they could transmit thoughts to each other through brainwaves. They also said that they would be willing to prove it.

The researchers, wary of deception, took precautions to make sure the two men were not cheating. Smith had his eyes bandaged and his ears plugged with cotton. He was then covered in heavy blankets. Amazingly, even though he was covered, Smith could repeat what the researchers said to Blackburn. He could even duplicate pictures that the scientists drew and gave to Blackburn. The only problem Smith and Blackburn ever encountered was when skeptical scientists monitored the experiment. Smith and Blackburn said that the skepticism hurt their powers. How did the two gentlemen manage to "mentally" transfer images to each other?

4) Occam's Razor is also called the "Rule of Parsimony". What does parsimony mean and how does it relate to Occam's Razor?

5) Mr. Credulous, who was born on June 17th, reads the astrology page each day. He is a strong believer in the accuracy of astrology because he claims that the horoscopes he reads always describe him accurately. Look at the sample horoscope below and try to decide why Mr. Credulous thinks that astrology is legitimate.

> *You are a sensitive individual---much more than people think. You feel that you are not totally understood by those around you. You are sometimes shy, but can be outgoing. Even though you make mistakes, you are a good person at heart. You are hard on yourself sometimes. You do not like it when people are mad at you.*

Occam's Razor
Einstein Level

1) Richard Feynman once said about UFO's: "I believe it is much more likely they are the result of irrational behavior by terrestrials than rational behavior by extraterrestrials." What did Richard Feynman mean by this statement?

2) Dr. Adams was a very strange doctor who lived in San Francisco in the early 1900's. He believed that each ailment had a vibration rate and that he could tell what was wrong with an individual just by tapping on his spine or stomach. Using this theory, Dr. Adams invented the "dynamizer". This machine could not only tell what ailed a person, but it could also tell his age, sex, and religion. The machine was so powerful that it could even take readings over the phone. Even with all its powers, this machine was primitive compared to Dr. Adam's final invention.

The "oscilloclast" could cure diseases through the power of vibrations. Because Dr. Adams believed that each disease had a specific vibration rate, all that was required of a doctor was to set the machine to that vibration level and the patient would be cured. Dr. Adams rented his machine to hundreds of doctors around the world and made millions of dollars by the time he died in 1924. There was something odd about the contract that doctors who rented his machine were required to sign. It said that they must never open the box that contained the working parts of the "oscilloclast". What do you think was inside the box?

3) A famous magician made a claim that he could mentally bend a spoon. How did he make it appear as though he was bending a spoon with only mental energy?

4) Several people own statues that appear to cry tears. Thousands of people visit these statues to witness this strange phenomenon. Use Occam's Razor to give a simple explanation for the crying statues.

5) Thousands of people are buying pyramid-shaped structures because they are told that pyramids have many mysterious powers. A farmer bought a pyramid to put over his well because it was contaminated with bacteria. One week later, the farmer came into the store that sold him the pyramid to thank them. The bacteria were no longer in his water supply -- the pyramid appeared to fix the problem. Provide a simple explanation for the disappearance of the bacteria in the farmer's water.

Chapter 4

Mistakes and Frustration are a Part of Learning and a Part of Life

Challenge and frustration are a part of learning and life. They should both be viewed as a normal part of the learning process.

How Nature Deals With Mistakes

If a deer is born without a natural fear of lions, that deer in all likelihood will not survive very long. Because it will not survive very long, the "I'm not scared of lion" genes will not get passed on to future generations of deer.

If a mouse is born without a natural fear of cats, that mouse in all likelihood will also not survive very long. Because it will not survive very long, the "I'm not scared of cats" genes will not get passed on to future generations of mice.

Nature does not tolerate serious lapses in judgment from people either. It is often said about nature that it is coldly indifferent. When two gentlemen decided that it would be a good idea to swim across the moat at the Brooklyn Zoo and harass the polar bears, nature punished them for their lapse in judgment. Their "wouldn't it be a good idea to harass the polar bears" genes did not get passed on to the next generation.

Discussion Questions:

1) If there is a herd of 100 deer that are hunted by lions, what do you think are the traits of the deer that survive?

2) What are the traits of the lions that survive?

3) It is said that "mistake-prone" genes often do not get passed on to future generations. What does this mean?

The Simple Mistake that Destroyed the Mars Climate Orbiter

The 125 million dollar Mars Orbiter was just completing its 10-month journey to Mars when NASA's mission navigation team in California sent commands to the spacecraft to move it into orbit around the Red Planet.

The Mars orbiter was a crucial part of a double mission to Mars that would have allowed scientists to learn much more about the history of the existence of water on the planet. From this information, scientists hoped to gain some insight into the possibility of life having once existed on Mars.

When the Mars Climate Orbiter was built, it was programmed to use metric units, so when the commands to fire its engine reached the spacecraft, it read the numbers as metric units. Unfortunately, because the mission navigation team calculated the maneuvering information in English units, the commands sent to the spacecraft forced it perilously close to the planet, where it almost certainly was destroyed.

When this mistake was studied, the most disturbing part was not the mistake itself ----- NASA expects many mistakes to be made during the process of spacecraft development. What bothered those who studied the destruction of the Mars Climate Orbiter was the failure of NASA's process of checks and balances to catch the error.

You want me to go what direction at what speed?????? Okay, whatever you say............I'm just a dumb machine and you're the smart engineers.

I use to get really upset over long division errors. That 125 million dollar mistake makes my mistakes seem trivial.

Discussion Questions:

1) Explain why sending a number in English units to a spacecraft that was programmed to accept metric units would cause it to crash.

2) On a scale of 1 to 10, where 1 is a very minor error and 10 is the most serious error, how would you rate the destruction of the Mars Climate Orbiter? Why?

3) Would you fire those responsible for the error? Why or why not?

The Overloaded Bridge

The Golden Gate Bridge is an engineering marvel. With a main span of almost 4200 feet, it is one of the longest and most elegant suspension bridges in the world.

In 1987, as part of a 50th anniversary celebration, the bridge was closed to traffic so pedestrians could celebrate by walking out onto the bridge. As dawn broke on the morning of the celebration, close to 1/2 million people showed up on each side of the bridge. When the mass of humanity met at the center, close to 1/4 million people were standing on the Golden Gate Bridge.

Even though the vehicular traffic that the bridge usually carried consisted of cars and trucks that weighed thousands of pounds each, the weight of 250,000 people was far more weight than the bridge had ever been subjected to. Several people who stood on the shore across from the bridge reported that the arch in the middle of the bridge flattened close to 10 feet.

In addition to the flattening of the arch, it was also reported that some of the hanger cables became slack as the weight of 1/4 million people stressed the structural components of the bridge.

Did the Golden Gate Bridge come close to collapsing that day in 1987? Probably not. But if the weight of 1/4 million people exposed a previously unknown structural flaw, a tragedy of unimaginable horror could have occurred.

It is indeed fortunate that the Golden Gate Bridge was able to hold the weight of so many people. It is an unfortunate fact that when structures are subjected to stress, unforeseen problems can and do arise.

They should have only allowed mice and rats on the bridge that day.

Discussion Questions:

1) Do you suppose that there were concerns about the amount of weight on the bridge before the celebration?

2) Estimate the amount of weight on the bridge that morning.

3) If the bridge did collapse, would you place any blame on the people who built the bridge?

The Apollo 13 -Triumph from Tragedy

The one event that astronauts fear the most is the sudden tremor or jolt that indicates that their craft has been hit by a meteor. If a spacecraft was hit by a meteor the size of a grain of sand, it would have devastating consequences. Almost instantly, the life-sustaining atmosphere within the spacecraft would be released into space, causing certain death for the astronauts.

Apollo 13 was in the third day of its mission to the moon and almost 200,000 miles from Earth, when the spacecraft was rocked by a terrifying jolt. Although it soon became apparent that the jolt was not caused by a meteor, the relief was short-lived.

As it turned out, the violent tremor was caused by the explosion of one of the two main oxygen tanks that provided not only much of the power to run the spacecraft, but more importantly, the oxygen to sustain life. During the next three and a half days, the astronauts and the engineers at Mission Control faced many life-threatening situations as they struggled to get the damaged spacecraft home.

Shortly after Apollo 13 parachuted to a safe landing in the Pacific Ocean, an inquiry began into how a catastrophic event such as an oxygen tank explosion could occur in a spacecraft with so many safety features.

After a thorough study, the cause of the explosion was determined. As so often happens, this near-tragedy was not caused by one mistake, but several small ones that by themselves wouldn't have caused a serious problem. Unfortunately, when these mistakes were added together, they lead to the catastrophic explosion of oxygen tank two and the near loss of the crew of Apollo 13.

Mistake #1:

The oxygen tanks that were used in Apollo 13 were kept at a temperature of -340°F. This temperature kept the oxygen in a slushy, nongaseous state, but still allowed for vaporization to occur so oxygen could flow to the spacecraft. Occasionally, the slushy oxygen would not vaporize properly, so fans and heaters were put into the oxygen tank. This allowed the astronauts to heat and stir the oxygen if more vaporization was needed.

Because of the explosive nature of pure oxygen, putting electrical equipment inside a tank of oxygen can be very dangerous. Because of this danger, the heaters had thermostats that would switch the electricity off if the temperature rose too high. In this case, the engineers designed the system to shut off if the temperature rose above 80°F.

The tank heaters were originally designed with a 28-volt power grid, but before construction began, this relatively delicate power grid started to worry the engineers who were in charge of designing the Apollo spacecraft. Because of their concerns, an order went out to change the power grid for the fans and heaters to a 65-volt system.

The contractors modified the entire tank electrical system to a 65-volt grid except for one small part. The original 28-volt thermostat switches were not replaced with 65-volt switches.

If you send current from a 65-volt system through me, I will fuse shut and then will not be able to shut off the current if the temperature starts to climb to dangerous levels.

This mistake alone probably wouldn't have caused a serious problem, but when more mistakes occurred, the stage was set for a catastrophic explosion.

Mistake 2:

When an oxygen tank needed to be removed from an Apollo spacecraft, it was not an easy procedure. A small crane was attached to the shelf that held the tank, and the shelf would then be slowly hoisted out of the spacecraft. During one of these procedures, one of the four bolts that held the shelf of oxygen tank two in place was inadvertently left fastened. While the tank was being lifted, it suddenly slipped off the crane and fell two inches back to its original position.

NASA accident procedures called for a thorough inspection of the tank when this mishap occurred. The inspection showed no damage, but in reality a small vent tube inside the tank had broken when it fell from the hoist.

Mistake 3:

In the days before the launch of a manned spacecraft, a full dress rehearsal is usually conducted. In addition to fully suiting the astronauts, the oxygen tanks are fully pressurized during these exercises. The rehearsal for Apollo 13 was uneventful, except for the fact that the technicians were unable to drain the oxygen from tank two when the rehearsal ended.

After looking at the history of oxygen tank two, the engineers quickly realized that when the tank was dropped a year and a half earlier, the fall must have damaged the vent tube that drains the tank. Because replacing tank two would have meant a month long delay, the engineers came up with what they thought was a simple and safe procedure. They would force the oxygen to vent by turning on the tank heater.

The engineers knew that the thermostat would shut off the heater if the temperature climbed above 80°, so they were not overly concerned about the possibility of the tank overheating. Even so, as an additional precaution, the engineers directed a technician to monitor the temperature of the tank in case the thermostat failed. If the temperature in the tank did exceed 80°, he would know that a problem had occurred with the thermostat and would immediately shut off the heater.

As the heaters started to heat oxygen tank two, the cumulative effects of several small mistakes converged into one serious error. As the temperature neared 80°, the 28-volt thermostat tried to open and shut off the electric current, but it was fused shut by the 65-volt current.

As the temperature rose to dangerous levels, the technician was the second line of defense. He should have easily spotted the high temperatures and shut the heaters off. Unfortunately, the thermometer that he was monitoring had an upper limit of 80°. While the temperature in tank two reached levels above 1000°, the temperature gauge told the technician that the tank was at a warm but safe 80°.

Just as the engineers had hoped, the heaters forced the liquid oxygen out of the tank. What the engineers didn't know was that while the oxygen was venting, the extreme heat also melted the teflon insulation that was around the wires which ran through the tank.

As Apollo 13 progressed on its mission to the moon, oxygen tank two had bare copper wires running through pure liquid oxygen. When Apollo 13 was almost 1/4 million miles from Earth, the fans were turned on to mix the contents of the tank. Because the wires within the tank were now bare, a spark ignited a fire and the resulting pressure blew the neck off the tank. Fortunately, the remaining oxygen tank remained intact, but because it shared plumbing with tank two, it began leaking its oxygen into space as the three astronauts scrambled for the safety of the lunar lander.

Even though Apollo 13 was a highly engineered machine with well-designed safety systems, it came close to being destroyed by very simple mistakes.

Discussion Questions:

1) Why would a very small meteor be so dangerous to a spacecraft?

2) Why was the oxygen kept at so cold a temperature?

3) Of the mistakes mentioned, what mistake do you think was the most serious?

It is Important to Experience
Failure and Frustration

An important study on failure was conducted by a University of California psychology professor named Salvatore Maddi.

When the telephone industry was deregulated, dramatic changes occurred in the lives of executives who worked for the phone industry. Salvatore Maddi's study involved following close to 500 mid-level executives at Illinois Bell as they dealt with the stress of changing job roles.

What Maddi discovered was shocking. Close to two thirds of the executives had a difficult time dealing with the changes that deregulation brought to their lives. What was even more interesting than the struggles that these executives went through, was the response of the remaining executives to the disruption in their lives. They not only adapted well to change, but many actually thrived under the same adverse conditions that caused almost two thirds of their colleagues to fall apart.

Maddi was intrigued by these different responses to challenge and frustration. When he probed further, he found that those executives who thrived during the stressful times had life experiences that were similar. All experienced challenge and frustration as children due to sickness, constant moving, the death of someone close to them, or other tough conditions.

Those executives who didn't fare well, typically had childhoods that were fairly stress-free. The challenges and frustrations that deregulation brought to their lives were foreign to them, so they had no built-in coping mechanisms to help them respond in positive ways.

Maddi reached the conclusion that when children experience challenge and frustration, the experience helps them develop resiliency. This resiliency enables them to deal with trying experiences as adults.

So what you are saying is that when I get problems like the ones that are at the end of this chapter, I should appreciate the fact that my mind feels like exploding.

Discussion Questions:

1) Why would adults who had stress-free childhoods have trouble coping with dramatic changes in their lives?

2) Maddi's study would seem to support exposing children to some challenge and frustration. Why would this be helpful?

3) Some children are so used to always getting the right answer that they are afraid to try anything that they might possibly get wrong. In other words, they lock themselves inside an intellectual box. Why is this not a good thing to do?   .

The Three Mile Island Accident

Almost nobody thought it would happen, but early one morning on March 28th, 1979, the Three Mile Island nuclear power plant came close to a catastrophic nuclear meltdown. A series of equipment failures that were compounded by human error, resulted in the exposure and partial meltdown of the nuclear reactor's core.

The nightmare started when a valve failed to close and water that was responsible for cooling the nuclear fuel began to drain. As the core's temperature steadily rose, the confused plant operators shut off the emergency cooling system. This action sent the core temperature rapidly climbing to over 4000° Fahrenheit. As the temperature climbed, the console in the plant started flashing hundreds of warning lights. To further add to the mounting chaos, warning sirens also began sounding.

If the core temperature continued to climb to 5200°, a meltdown would occur. If that happened, the nuclear core would go through the concrete floor of the plant and into the ground water. A deadly radioactive steam-cloud would then blow over the nearby city of Harrisburg.

As the core of the reactor continued to approach meltdown temperatures, the operators mistakenly believed that it was covered in water. The plant operators didn't fully appreciate the seriousness of the situation until radioactive gases started showing up inside the plant. At that point, a general emergency was declared.

State government officials knew that the public needed to be reassured, so Lieutenant Governor William Scranton held a news conference where he informed the public that no radiation had escaped from the facility and that Metropolitan Edison had the situation under control.

A few minutes after he spoke, Scranton learned that radiation had in fact been released from the plant. This incident made him realize that the information that was coming from Metropolitan Edison was not as reliable as it should have been.

If there is any time you need accurate information, it is when there has been a nuclear accident.

Faced with uncertainty and the mounting seriousness of the situation, Governor Thornburgh had no alternative but to recommend that pregnant women and children begin an evacuation.

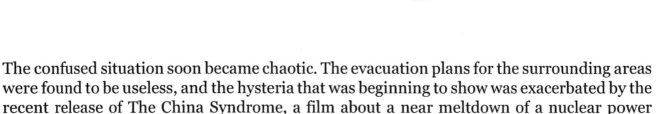

The confused situation soon became chaotic. The evacuation plans for the surrounding areas were found to be useless, and the hysteria that was beginning to show was exacerbated by the recent release of The China Syndrome, a film about a near meltdown of a nuclear power plant.

The situation at the plant continued to worsen. People in the control room were now talking through respirators, which made communication very difficult. In addition, only two outside phone lines connected the control room with the outside world. Meanwhile, the designers of the reactor (Babcock and Wilcox) were desperately trying to reach the operators, but because of the lack of phones lines, they were unable to contact them. After five frustrating hours, Babcock and Wilcox engineers were finally able to get a message to the control room to "get water moving through the core!"

When operators did restart the pumps, the temperature inside the reactor immediately dropped and many thought the crisis was over.

Two days later, another serious situation arose when a hydrogen bubble formed above the core of the reactor. It was feared that this bubble could prevent cooling of the core and lead to a meltdown. In addition, there was a very real possibility that the hydrogen could mix with oxygen and explode. After several tension filled days, the situation stabilized and the hydrogen was drained from the system. Four weeks later, the reactor was shut down.

Several years later, the core of the reactor was examined by a robotic camera. What the engineers found was both shocking and disturbing. There was a meltdown at Three Mile Island! A significant part of the core had in fact been melted and destroyed by the high temperatures that occurred when a leak led to an ill-advised shutdown of the emergency cooling system during those harrowing days in 1979.

The Three Mile Island accident started with a malfunctioning valve. When the accident was analyzed, it was discovered that there had been a problem with this valve eleven other times at other plants, but no warnings or corrective measures had been undertaken.

Other equally disturbing situations came to light after the accident was studied. The evacuation plan was so inadequate that it even directed residents of two different communities to travel toward each other over the same bridge.

It also became clear that the people in the area surrounding Three Mile Island were not honestly informed as to the seriousness of the situation.

Discussion Questions:

1) Why were the managers at Three Mile Island hesitant to declare an emergency much earlier?

2) How serious a mistake would you consider the poor evacuation plans?

3) Would it have been better for the public officials to have been very honest, even if it caused a panic?

The Hubble Telescope

Planning for the Hubble telescope started in the 1960's. It was an expensive project, but scientists hoped the telescope would gather information that would help answer many questions about the universe.

How big is the universe? How old is it?

I hope the Hubble telescope will answer those questions by finding out the speed of the expansion of the universe.

After years of planning and construction, the Space Shuttle Discovery took the 2 billion dollar telescope into space. Shortly after it was unloaded from the storage bay of the shuttle, many problems developed. Half of the Hubble's gyroscopes failed, sensors failed, computer memory failed, and its two solar panels were moving so much that they were in danger of breaking. To make matters even worse, scientists soon discovered that the images Hubble was sending to Earth were blurry!

After the images were mathematically studied, scientists discovered that Hubble's 8-foot mirror was not focusing light rays properly. It was supposed to take starlight and focus it to a single focal point, but because the mirror was built incorrectly, it could not focus the light. After a short investigation, the problem with the mirror was traced back to a measuring device that was used when the mirror was made.

The measuring device was called a reflective null corrector. It was out of adjustment by one millimeter.

That doesn't seem like a very large error. One millimeter is very small so how could that cause fuzzy images?

A one millimeter mistake in optical equipment is very large. Most optical errors are in the range of a 20th to a 50th of a millimeter. This error was so large that it was almost like using a ruler to build a house and then finding out, after the house was done, that what you thought was a 12 inch ruler was really only 11 inches long.

Fortunately, the Hubble was made in such a way that repairs were fairly easy to make. After the repairs were completed, Hubble was able to send astonishing images to Earth. Because of the extraordinary power of the Hubble telescope, we have been able to witness the birth of stars, watch comet fragments crash into Jupiter, see evidence of the existence of massive black holes at the center of galaxies, and increase our knowledge about the age of the universe.

Discussion Questions:

1) Why does looking at distant objects in our universe help us tell how old the universe is?

2) When we are looking at a star that is 1000 light years away, we are looking back in time. Explain why this is true.

3) Is a light year a measure of time or distance?

Mistakes and Frustration
Level 1

1) How many square inches are in a square foot?

2) What is the difference between mass and weight?

3) What is $8 \div 1/2$

4) What is the median of these five salaries? $100,000; $25,000; $50,000; $200,000; $100,000

5) What is the difference between speed and acceleration?

6) Write the number five hundred and two hundredths.

7) What is the difference between a foot, a square foot, and a cubic foot?

8) Which number is larger? .0100 or .009888?

9) There are 1000 meters in a kilometer. How many kilometers are in a meter?

10) What is larger, n or $2n$? Why?

Mistakes and Frustration
Level 2

1) How many cubic inches are in a cubic foot?

2) What is the difference between mass and volume?

3) Mike measured the diameter and circumference of a circle and then made a fraction using the diameter as the denominator and the circumference as the numerator. After that he turned the fraction into a decimal and rounded to the nearest hundredth. What number did he get?

4) What is .5% of $400?

5) What is $10^{89} \div 10^{88}$?

6) Is the number one a prime number?

7) Is n always bigger than $-n$?

8) Temperatures can rise to very high levels, even reaching above 25,000,000 degrees. How low can temperatures go? Why?

9) What is the difference between a bacterial illness and a viral illness?

10) If there is a giant piece of chocolate that weighs 250 pounds, how many 4/5 pound pieces can be cut from it?

Mistakes and Frustration
Einstein Level (1)

1) Americans use pounds when they measure weight. What is the metric system's unit of weight?

2) Americans use pounds to measure weight. What is the American unit for mass?

3) A store had a 50% off sale. On Sunday the ad said that from 10:00 A.M. to 2:00 P.M. an additional 25% can be taken off the price. At 12:00 noon on Sunday, Natalie bought a coat that had a regular price of $100. What did she pay?

4) On Richard's first day as a sales clerk, he charged a 7% sales tax when he should have charged 6%. The total amount of money he collected was $856. What should he have collected if he charged the correct tax?

5) $5^3 = 125$ $5^2 = 25$ What is 5 to the first power? What is 5 to the zero power?

6) Natalie was asked to find the prime factors of 60. She wrote 2 x 2 x 3 x 5. Natalie's answer was marked wrong. Why?

7) If an American dollar is equal to $1.32 Canadian, what is a Canadian dollar worth in American dollars?

8) It takes Bill two hours to paint a fence. It takes Steve four hours to paint the same fence. How long will it take them to paint the fence if they work together?

9) There is a temperature that is the same for both Fahrenheit and Centigrade. What is that temperature?

10) What fraction is halfway between 3/16 and 3/8?

Mistakes and Frustration
Einstein Level (2)

1) Eight farmers each had eight farms that each had eight dogs that each had eight puppies that each had eight fleas. How many total legs are there?

2) Ted had scores of 78.5%, 40%, 98%, 98%, and 100%. If Ted wants a final average of 86% and he has one more test to take, what score must he get on his final test?

3) A square piece of paper is folded in half vertically and then folded again vertically. If the perimeter of the resulting figure is 60 inches, what is the area of the original square?

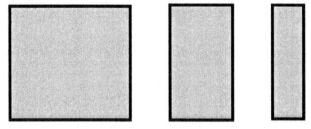

4) 12 college students bought enough food for an 8-day canoe trip. If 6 more students decide to go on the trip, how long will the food last?

5) If 6 days before the day before yesterday is Thursday, what day of the week is yesterday?

6) How many squares of any size are in the picture? Hint: Don't count! Find a pattern.

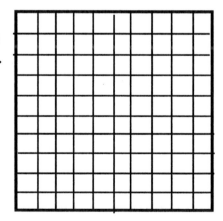

7) A 5" cube is made up of all 1" cubes. The outside of the 5" cube is painted. What percent of the faces of the 1" cubes are painted?

8) What is the ones digit in the number 8^{49}

9) A blue light flashes every 8 minutes and a green light flashes every 9 minutes. If they both flash at 3:12, when is the next time they will flash together?

Mistakes and Frustration
Einstein Level (3)

1) Rachel runs 7.2 miles in 42 minutes. If she runs at the same pace, how long will it take her to run an eight mile race?

2) The sum of three numbers is 301 and their ratio is 3:11:29. What is the largest number?

3) D is the midpoint of line AE and is 7/8 of the way from A to G. If the length of AD is 14," what is the length of GE?

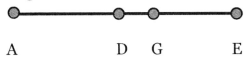

    A   D   G   E

4) Three pencils and two notebooks cost $3.75. Three notebooks and two pencils cost $3.65. What is the cost of a pencil?

5) Knives come in packages of 10. Forks come in packages of 8 and spoons come in packages of 6. If Nancy wants to end up with the same number of knives, forks, and spoons and doesn't want to throw any away, how many forks will she end up buying?

6) What 2-digit number is a perfect square and a perfect cube?

7) Sara was planning to build a cement driveway that was 27 feet long by 9 feet wide by 6 inches thick. If cement cost $35 a cubic yard, what will the cost of the cement be?

8) What is the next term in the following sequence?   1, 8, 27, 64

9) Jacob has a favorite recipe that makes a 3" thick cake when he uses a 10" by 13" pan. Unfortunately, the last time he made a cake, he dropped the pan and broke it. If Jacob uses the same recipe, how thick will the cake be if he uses a circular pan with a 9" diameter? (Round to the nearest tenth of an inch.)

Mistakes and Frustration
Einstein Level (4)

1) Write $\dfrac{.0005}{.00005}$ as a percent.

2) 1/8 of a brick plus 50 pounds is equal to 1/4 of a brick plus 38 pounds. How much does a brick weigh?

3) A clock loses 12 minutes every hour. If the broken clock and the normal clock are both set to 10:00 A.M., what time will the normal clock read when the broken clock reads 12:00 noon?

4) It took Bill 3 hours and 48 minutes to travel to his mother's house. If his average speed was 36 mph., how far away is his mother's house? (Round to the nearest mile)

5) The average of $\dfrac{1}{2}+\dfrac{1}{3}+\dfrac{1}{4}+\dfrac{1}{6}+\dfrac{n}{12}=\dfrac{1}{3}$ What must n equal?

6 If you can run 880 feet per minute, how many days would it take you to run 4560 miles?

7) Light travels at a speed of 186,000 miles per second. What distance is a light year?

8) There are 1000 students at Jones Junior High School. 55% of the students are girls and 40% of the girls like basketball. 10% of the girls who like basketball are 6 feet tall and 50% of these girls are A-students. How many 6 foot tall girls who like basketball and are A-students are there at Jones Junior High School?

9) Simplify: $\dfrac{\dfrac{1}{3}-\dfrac{1}{4}}{\dfrac{1}{3}+\dfrac{1}{4}}$

Mistakes and Frustration
Einstein Level (5)

1) Jill walked 1/4 of the way home at 4 mph. She found a bike and went another 1/4 of the way home at 16 mph. A tire on her bike went flat so she rode the rest of the way home in a car at a speed of 48 mph. What was Jill's average speed on her trip home? (Round your answer to the nearest tenth.)

2) Isaac rolled 3 dice. What is the probability that he would roll three of a kind?

3) A broken clock loses 4 minutes every hour. If the broken clock and the normal clock were both set for 12:00, what time does the normal clock say when the broken one reads 2:06?

4) Simplify: $7\dfrac{2}{5\dfrac{1}{5\frac{2}{3}}}$

5) Isaac had a pile of quarters, dimes and nickels that had a value of $18. If the number of quarters was three times the number of dimes and the number of dimes was seven times the number of nickels, how many dimes does Isaac have?

6 A piece of paper $8\dfrac{1}{2}$ inches by 11 inches is rolled in two different ways to make two different cylinders. How much bigger is the volume of the big cylinder when compared to the small cylinder? (Round to the nearest whole number.)

7) What is the 81st term in this series? 1, 16, 81, 256

8) 15% of the student body at Einstein Middle School are left-handed. 20% of the left-handed students play soccer and of those, 5% play goalie. What percent of the student body are left-handed soccer goalies?

9) The number .000337 is a million times larger than what number?

Chapter 5

It is Important to Keep an Open Mind

Many of the greatest ideas and discoveries of the last 500 years were ridiculed and attacked when they were initially presented. The medical advances that were delayed because of this attitude are tragic examples of the implications of a failure to keep an open mind.

On March 11, 1878, the president of the French Academy of Sciences was presenting Thomas Edison's new invention to fellow scientists.

How dare you accuse me of not having an open mind.

The man who was trying to strangle the president of the French Academy of Sciences was Jean Bouillaud. He was a fellow scientist who thought it was impossible to duplicate the noble human voice. I think it is safe to say that he wasn't open to new ideas.

When you look back on this story, it is a little bit humorous. The stories you will be reading in chapter 5 are not funny. Thousands of people died or suffered because people who were in positions of power did not keep their minds open to new ideas.

Elizabeth Kenny and Polio

"I have encountered many 'open minds' in the medical profession in many parts of the world, and I have learned to proceed with caution when I come upon one now. Some minds remain open long enough for the truth not only to enter, but to pass on through by way of a ready exit without pausing anywhere along the route."
- Sister Elizabeth Kenny -

In 1911, Elizabeth Kenny worked alone in the Australian Outback trying to bring medical care to the homesteaders who lived in the desolate area. She was poorly trained as a nurse, but tried her best to provide medical care to a group of people she had come to love dearly.

Early that summer, Sister Kenny was called to the home of a family she knew well. Her favorite of the family, a little girl named Amy, was lying upon a cot in such clear and obvious agony from the effects of polio that Sister Kenny thought her heart would break. The little girl's muscles were in a strange kind of spasm that caused the child such pain and discomfort that sister Kenny knew that she had to act quickly.

Because Sister Kenny had never encountered a case of polio, she was anxious to find the proper treatment as soon as possible so she could alleviate the poor girl's suffering. She telegraphed a family friend who was a doctor asking for advice. Later in the day, Sister Kenny received a reply that gave her very little encouragement. " **INFANTILE PARALYSIS. NO KNOWN TREATMENT. DO THE BEST YOU CAN WITH THE SYMPTOMS PRESENTING THEMSELVES.**"

As it turned out, it was fortunate that the doctor did not tell Sister Kenny what the standard treatment for polio was. If he had, the child likely would have had splints attached to her limbs that were in spasms. Because she did not know that immobilizing the limbs was the suggested treatment for polio, Sister Kenny used her intuition. Because the muscles were very tense, Sister Kenny thought a reasonable treatment would be the application of hot compresses to the arms and legs to encourage relaxation.

Over time, the hot compresses helped end the agonizing spasms. But now, the little girl had limbs that were so affected by the polio virus that they could barely move. Using her intuition again, Sister Kenny theorized that the limbs needed reeducation and increased blood flow. She immediately embarked on a regimen of motion therapy and massage.

The results of Sister Kenny's innovative treatment were dramatic. Her method worked so well that Amy and many other children who were treated in this fashion ended up walking again. When doctors saw the results of Sister Kenny's work -- ambulatory children who often showed little or no signs of their bout with polio -- their reaction was very surprising.

Instead of trying to find out about this technique and then studying it further, they flatly rejected it by saying that a mistake was obviously made. Sister Kenny must have been in error when she determined that these children had polio.

It is obviously a mistake. If these children had polio, they would be crippled.

Sister Kenny spent 30 years trying to make the medical establishment understand that they were treating polio in a way that almost always ensured that the patients would not recover the use of their affected limbs. Sister Kenny's efforts were rewarded with rejection, ridicule, and ostracism.

Because doctors had "learned" that the correct treatment for polio was immobilization, their minds were rigidly closed to new ideas.

It almost seems like Sister Kenny's lack of knowledge was a good thing! She allowed her mind to think about the disease and then come to a conclusion about treatments that made sense.

In 1941, more than 30 years after Sister Kenny first discovered a treatment for the devastating affects of polio, the medical establishment finally recognized that hot compresses and motion therapy should replace immobilization as the standard treatment for polio. An editorial from the journal of the American Medical Association showed the shift in thinking that occurred at that time:

> *"It is of the opinion of this Committee on Research for the Prevention and Treatment of After-Effects of the National Foundation for Infantile Paralysis, after a study of the report of the workers at the University of Minnesota, that during the early stages of infantile paralysis the length of time during which pain, tenderness, and spasm are present is greatly reduced and contractures caused by muscles shortening during this period are prevented by the Kenny method."*

Many of medicines greatest discoveries were initially met with doubt and ridicule before they eventually gained general acceptance. Because Sister Kenny's treatment was ignored for 30 years, tens of thousands of people suffered from the devastating consequences of an ill-advised medical treatment.

Discussion Questions:

1) If Sister Kenny was a male nurse or a doctor, do you think it would have helped get her ideas accepted by the doctors?

2) Do you think that Sister Kenny would have tried hot compresses and motion therapy if she knew the standard treatment for polio?

3) Why do you think the standard treatment for polio in the early 1900's was casting the limbs and then putting them in braces?

Dr. Semmelweis and Puerperal Fever

The year was 1847 and Dr. Semmelweis was faced with a serious problem. A significant number of women who gave birth at his hospital entered the hospital healthy, gave birth, and then died several days later of puerperal fever. What was very perplexing to Dr. Semmelweis was the fact that in Ward One, where doctors and medical students delivered babies, the death rate was close to 30%. In Ward Two, where midwives were in charge of the birth procedures, the death rate was 3%. The difference in the death rates between the two wards made it clear that the doctors were doing something that was causing women to catch puerperal fever. The evidence against the doctors was strengthened even further when the hospital conducted an experiment that required the doctors and midwives to switch wards. The results were shocking! The death rates followed each group of medical practitioners. The death rate at Ward One experienced a significant drop, while the death rate in Ward Two jumped dramatically.

Dr. Semmelweis was deeply touched by the suffering and death that was occurring at his hospital. At the time of this great tragedy, the concept of germs causing disease had not yet been established, so Dr. Semmelweis racked his brain trying to find a possible cause for this strange disease that was killing women who gave birth.

Around this time, a colleague of Dr. Semmelweis's cut his finger during an autopsy on a woman who had died of puerperal fever. This cut soon became infected and the doctor ended up dying of the same disease as the woman -- puerperal fever. Dr. Semmelweis knew that doctors went directly from autopsies to birthing rooms, so he took this strange occurrence as an indication that something was being carried from the dead in the autopsy room to the women giving birth. He thought that if the doctors washed this "death" off their hands, it might prevent puerperal fever.

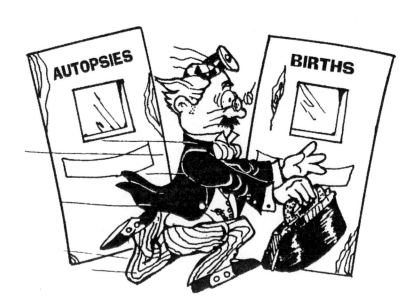

Dr. Semmelweis directed every doctor he supervised to wash their hands in a chlorinated solution before they assisted at births or examined pregnant women. Even though many doctors resisted this new directive, most complied.

The results of Dr. Semmelweis's mandatory handwashing experiment were dramatic. The death rate dropped to less than 2%. Dr. Semmelweis was confident that he had found the cause of puerperal fever and an easy way to prevent it, but his hope that the medical establishment would adopt handwashing was quickly dashed.

The director of the hospital felt that Dr. Semmelweis had overstepped his authority and became hostile to the idea of handwashing. As the months turned into years, it became increasingly clear that handwashing saved lives, but still Dr. Semmelweis was criticized and ridiculed by his colleagues. He was eventually fired.

After an extensive, but futile effort to convince the medical establishment of the benefits of handwashing, the stress of the criticism and the thousands of needless deaths caused Dr. Semmelweis to have a mental breakdown. He died in the summer of 1865, a broken and bitter man.

The medical establishment ignored clear evidence that would have saved thousands of lives. Dr. Semmelweis's idea about cleanliness was so new that very few people were willing to look at it with an open mind. Ironically, a short time after Dr. Semmelweis died, Joseph Lister started using carbolic acid during surgery to prevent infections. Joseph Lister often said that without Dr. Semmelweis, his achievements would have been nothing.

Discussion Questions:

1) Why do you think the other doctors rejected Dr. Semmelweis's theory concerning the cause of puerperal fever?

2) Were midwives at Dr. Semmelweis's hospital more knowledgable than doctors as to the causes of disease?

3) What more could Dr. Semmelweis have done to convince the other doctors that handwashing could prevent fatal diseases in women who were giving birth?

Dr. Goldberger and Pellagra

Pellagra was a terrifying and painful disease that was contracted by over 100,000 Americans each year during the early 1900's. It started as a hardening of the skin and then progressed to mental confusion, hallucinations and finally insanity. Because the disease was thought to be highly contagious and because it caused agony and mental confusion, people with pellagra were typically feared and avoided.

A strange characteristic of pellagra was that it usually was limited to orphanages, prisons and asylums. Because of the living conditions in these three locations, the medical establishment made the logical conclusion that pellagra was caused by an unsanitary lifestyle.

Dr. Goldberger, who was very interested in finding the cause of pellagra, did not share the view of the medical establishment that the disease was caused by microbes. He based his opinion on the fact that even though hundreds of patients at several insane asylums had pellagra, no nurses, guards or other employees contracted the disease. Because the workers often had close contact with the patients, Dr. Goldberger theorized that if pellagra was truly caused by microbes, then at least some of the employees would have contracted the disease.

After considering several other causes, Dr. Goldberger focused on the diet of the individuals who were suffering from pellagra. What he found shocked him. While workers had a diet that consisted of vegetables, eggs, fresh meat, and milk, the patients were forced to subsist on a sparse diet that consisted of a small amount of meat, cornmeal and molasses.

As all good scientists do, Dr. Goldberger put his theory to a rigid test. He used the scientific method to determine whether food played a role in the development of the horrible disease.

If his experiment proved that diet had a role in causing this disease, the medical establishment certainly couldn't ignore his results as they did with Sister Kenny and Dr. Semmelweis.

I wish they didn't ignore his results, but they did. Unfortunately, sometimes it is very hard to keep an open mind. The rest of the story is very sad.

Dr. Goldberger tried his experiment at an orphanage where a large number of children were suffering from pellagra. Dr. Goldberger changed the diet of the children by adding dairy products, beans, and fresh meat to their daily meals. He also made sure that no other conditions changed because he wanted to make sure that any improvements in the health of the children could be attributed to the modification of their diet.

After only a few weeks, the health of the children changed dramatically. The symptoms of pellagra began to fade, and even those children who weeks earlier seemed sure to die, began to recover. Because of the stunning evidence that Dr. Goldberger gathered, he was eager to share his discovery with the world. Unfortunately, the thinking of the time was so rigid that this dramatic evidence of the causal agent for pellagra was dismissed as meaningless.

As hundreds of people continued to die, Dr. Goldberger was determined to provide even more proof that pellagra was caused by a poor diet and not by microbes. Even though the medical establishment disagreed, Dr. Goldberger felt confident that he had proven that a better diet could cure pellagra. In an effort to provide more proof for his theory, he decided to try and induce pellagra in healthy individuals by placing them on a poor diet.

Dr. Goldberger was able to find volunteers for his experiment at a prison that offered full pardons for those who agreed to participate. He immediately placed them on a severely restricted diet that he was sure would eventually cause pellagra.

Just as he predicted, the prisoners started to develop symptoms of pellagra after several months on this sparse diet.

Surely the world would have to listen to Dr. Goldberger's theory now, wouldn't they? He showed that he could not only cure pellagra, but also induce it.

For more than 20 years, the government and the medical establishment ignored Dr. Goldberger's research. Hundreds of thousands of people died needlessly because people in positions to make changes believed what they wanted to believe and ignored what the scientific method told them. In the mid 1930's, a deficiency of a B vitamin called niacin was found to be the cause of pellagra. Dr. Goldberger was finally vindicated!

Discussion Questions:

1) Do you think it was ethical to do this experiment on prisoners?

2) Most doctors thought the cause of pellagra was bacterial. How do you think they explained the fact that employees at prisons, orphanages, and insane asylums did not get pellagra?

3) Goldberger used prisoners in his experiment. Even if people from outside the prison were willing to go on the restricted diet, it is unlikely that Dr. Goldberger would have used them. Why?

The Shocking Discovery About Ulcers

Up until a few years ago, doctors thought that ulcers were caused by too much stress, eating too many spicy foods, or drinking acidic beverages.

A dramatic shift has recently occurred in the thinking on what causes ulcers. It all started in 1983 when two Australian doctors -- Dr. Warren and Dr. Marshall -- discovered a certain kind of bacteria (H. pylori) in the stomachs of people with ulcers. This led them to think that ulcers might possibly be caused by bacteria and not stress or excess stomach acid.

Dr. Marshall thought that his evidence that bacteria caused ulcers was so convincing that the medical establishment would quickly accept his hypothesis. He was wrong.

Like Elizabeth Kenny, Dr. Marshall was young and did not yet have a strong belief system as to what caused ulcers. His mind was open to the possibility that there was another explanation. Many of his fellow doctors, who were experts in the field of gastroenterology, had very entrenched beliefs that left their minds almost completely closed to new ideas.

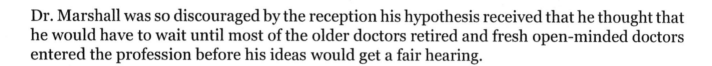

Dr. Marshall was so discouraged by the reception his hypothesis received that he thought that he would have to wait until most of the older doctors retired and fresh open-minded doctors entered the profession before his ideas would get a fair hearing.

Fortunately for Dr. Marshall, the medical establishment slowly started to accept his theory. After several medical researchers came to the same conclusion as Dr. Marshall, more and more doctors started to change their long-established beliefs to align with Dr. Marshall's.

> The medical establishment abandoned an incorrect belief more quickly than in the cases of polio, puerperal fever, and pellagra, but the hostile reaction to Dr. Marshall's new idea is still very disturbing. Many of the greatest ideas and discoveries of the last 500 years were ridiculed and attacked when they were initially presented. The medical advances that were delayed because of this attitude are tragic examples of the implications of a failure to keep an open mind.

Discussion Questions:

1) Why did Dr. Marshall think that his theory would not be accepted until the older doctors retired?

2) If Dr. Marshall tested 500 patients with ulcers and he found that they all had the H. pylori bacteria in their stomachs, would that alone be very strong evidence for his theory that the bacteria caused ulcers?

3) Do you think that Dr. Marshall did the right thing when he purposely took H. pylori to see if it would give him an ulcer?

Open Mind
Level 1

1) The people who developed the_____, initially did not get an enthusiastic reception from the scientific community. "The wheels will turn, but it won't move." After it did move, the naysayers in the scientific community said that the speed would cause brain concussions in passengers. Other scientists said that anyone who looked at the moving_____would get dizzy and faint. Several scientists even suggested that if the_____was built, a fence should be built along its route so that people would not be harmed by looking at the moving_____. What invention were people so nervous about?

2) In 1781, Francois Blanchard announced his design for a_____. A famous French astronomer dismissed his idea saying: "It is entirely impossible for man to rise into the air and float there. Only a fool would expect such a thing would happen." What did Blanchard design?

3) In the early 1800's, the French Academy of Sciences asked one of its members named Lavoisier to study a series of strange occurrences that had been talked about for hundreds of years. When Lavoisier finished his study, he declared confidently that "there was no such thing as_____. These objects must have been vomited by the Earth." What objects did Lavoisier study?

4) What invention prompted this statement: "There will not be any demand for them. I can see that some institutions might buy one or two, but other than that, I just don't see any interest."

5) When Remington first started selling his machine that could write, many people believed there were no practical uses for it. What did Remington invent?

Open Mind
Level 2

1) When Dr. William Harvey presented his theory on_____, his colleagues were so vicious with their criticism that most of his patients abandoned him. In addition to this theory, Harvey also proved that the liver was not the central organ of the vascular system.

2) One of the greatest medical breakthroughs of all time was the work of Edward Jenner. When he first presented his theory on_____his colleagues not only dismissed his idea, but they wrote many articles that attacked and ridiculed him. One doctor even made up a story about a boy who was inoculated with Jenner's "animal juices" and later began to walk on all fours and moo like a cow.

3)_____was a famous scientist who was one of the first people who realized that the Earth was not the center of the universe. His reward for this discovery was constant persecution. In 1633 he was given a sentence of life imprisonment. (The sentence was commuted to permanent house arrest.)

4) Nikola Tesla was a pioneer in this field. A well-known Austrian physicist wrote an article that was very critical of Mr. Tesla's ideas. In this article, the physicist expressed outrage that Tesla would dare claim that he could eventually transmit speech across thousands of miles or transmit music by electrical waves. So someday "we sit comfortably in an armchair, take the small receiving apparatus into our hand, switch it on and hear an opera sung at an immeasurable distance! This alone shows what an impractical, nay dangerous dreamer this so-called scientist is." (*History of Stupidity*) What field did Nikola Tesla pioneer?

5) The rights to this invention were offered to Western Union, but they saw it as little more than a toy.

Open Mind
Einstein Level

1) Luigi Galvani was a well-known physiologist who made several important discoveries about the effects of electricity on animal nerves and muscles. Galvani's colleagues at Bologna University did not think much of his theories and mockingly called him the "dancing master of frogs". Why was Galvani called the "dancing master of frogs"?

2) Dr. John Gorrie invented an innovative machine in the mid 1800's. Even though his machine eventually ended up making billions of dollars for others, Gorrie died a poor and bitter man because he couldn't find enough people willing to invest in his machine. Gorrie's invention made it possible to have cold drinks all year round. What did Gorrie invent?

3) In 1783, the Marquis Claude Francois Doothee de Jouffroyd'abbans invented a boat that allowed speedy travel without wind. When asked for comment from the Paris Academy, they said that the invention was not worth spending any money on. What was this "worthless" invention?

4) In 1915, A German meteorologist named Alfred Wegener looked at maps of the world and put forward a theory about what the world looked like hundreds of millions of years ago. His colleagues laughed at his theory, but it is accepted today. What was Wegener's theory?

5) Dr. _____ had a small engine placed in his operating room that sprayed a fine mist of carbolic acid during all operations that he conducted. He also poured carbolic acid into wounds and onto bandages. He also would not enter the operating room without scrubbing his hands with a disinfectant. His fellow doctors would laugh at him and make disparaging remarks behind his back. Dr. _____ got the last laugh when his surgical mortality rate dropped from 45% to 15%. When a Munich hospital adopted his methods during the Franco-Prussian war, their incidence of gangrene went from 80% to 0% in only one week. Who was this doctor?

Chapter 6

It is Important to Maintain a Healthy Skepticism

It is important to maintain a healthy skepticism. You must not have a mind that is so open that your brains fall out.

And I see snow storms hitting New York, a dry spell in Arizona..........

In addition to knowing almost everything about me, the psychic made some predictions that I am going to put on my refrigerator and see if they come true. I am so impressed with her skills that I have no doubt most of her predictions will be correct. She is incredible!

1) *An American politician will be involved in a scandal.*

2) *An actor will take a stand on a political issue.*

3) *A big name in entertainment will die unexpectedly.*

4) *There will be a violent conflict in Asia or Africa.*

5) *There will be a devastating earthquake somewhere in the world.*

6) *There will be a new controversial television show or movie that will be popular, but many will say it is inappropriate.*

7) *A well-known actress will get divorced.*

We mentioned in the last chapter that it is important to keep an open mind, but you can't go around believing everything. In the cartoon, all that is at stake is $4.95 per minute, but as you will see in these next stories, much more can be at stake when people don't use critical thinking.

My psychic is only $3.95 a minute, does that mean that she is less accurate?

Although it is very important to keep an open mind, we must close our minds to some things. When we do close our minds, it must be done carefully and thoughtfully.

Psychic Surgery

Seriously ill patients who have little or no hope of recovery, sometimes become so desperate for a cure that they turn to unscrupulous people who promise to heal them.

Even though medical doctors have given you no chance of recovery, I have the ability to reach into your body and tear out your tumor. I ask only two things of you. The first is $10,000 dollars in cash, and the second is to suspend your critical thinking for a few days.

Psychic surgery is one of the "healing procedures" practiced by people who take advantage of the desperately ill. Those who practice this kind of "surgery" claim to be able to remove tumors without scalpels or anesthesia. Even more amazing is the fact that after the "surgery" is completed, no skin wound exists--even though the surgeon is claiming that he reached inside the patient and removed a tumor.

In reality, practitioners of psychic surgery use sleight of hand to give the illusion that surgery is being performed. They may use false thumbs or fingers to hide animal blood and tissue. As the "surgery" progresses, a patient sees blood and tissue appear as tumors and diseased tissue are supposedly removed from the body. After the surgery is completed, the patients usually go home much poorer, but still seriously ill.

Pay no attention to my gigantic thumb or the chicken organs I have hidden behind my back.

The American Cancer Society has concluded that, "All demonstrations to date of psychic surgery have been done by various forms of trickery." When you have a healthy skepticism, Occam's Razor is a tool that can help you judge things such as psychic surgery.

When I am trying to decide whether a person really has the ability to reach into a body and pull out tumors, Occam's Razor tells me that it is much more likely that trickery is involved.

Discussion questions:

1) Why do people go to these "doctors" who are obviously involved in trickery?

2) Many people, after spending thousands of dollars on sham "surgery", report that they feel much better. How do you explain this apparent improvement in their condition?

3) What are some things you might say to someone who is considering this type of surgery to try and discourage them?

The Invention Company

In the 1980's and the 1990's, a commercial for an invention company would often play on late night television that asked inventors to send in their ideas for analysis. The company claimed that they would study the ideas, and if they were good, help people market their inventions.

Send your invention to us. We don't care if its large or small, simple or complex, we know how to market inventions. We might make you very rich.

I have an idea for a new kind of mousetrap. It has a 3-second delay after the cheese is touched, and the metal bar has thick padding.

When inventors would send their ideas to the company, they were usually told that their inventions were very promising. Shortly after they were given the good news about their idea, the invention company asked the inventors for money to help market their product. A man in Iowa, who we will call Mr. Cynac (he did not want his real name used), had several excited friends who sent hundreds of dollars to the invention company because they were told that projected earnings from their inventions were over one million dollars.

Mr. Cynac was very concerned that the invention company was taking advantage of people, so he devised a plan to see if the company could be trusted. He decided to send in plans for the most ridiculous and worthless invention he could think of. If the company said that his invention was a good idea, then he would know that it was a company that shouldn't be trusted.

After considering several ideas, Mr. Cynac decided to submit plans for a spring-loaded fork that harpooned food when a button was pushed. The "powerfork" was the perfect ridiculous invention because it was mechanical, expensive, and of absolutely no use to anyone. If the invention company said that the powerfork was a good invention, then Mr. Cynac would be sure that they were a company that was not being honest with people.

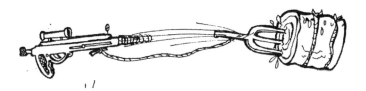

Mr. Cynac, your invention is the first one our entire board has been excited about! We are ready to move on this.

Two weeks after submitting his invention, Mr. Cynac received a letter from the invention company with high praise for his "powerfork". They told him that the earnings potential for his invention exceeded ten million dollars and that they could start marketing the "powerfork" as soon as they received a check for $495.

The invention company called Mr. Cynac repeatedly as it attempted to convince him to mail them $495. Even though the "powerfork" was a worthless invention, they praised him over and over about how smart he was. To stop their calling, Mr. Cynac told them that his invention was too dangerous because his mother had an accident with the "powerfork" and harpooned herself in the neck. For that reason, Mr. Cynac said, he was not interested in having his invention built and sold. He thought that would stop the calling, but it didn't. It wasn't until Mr. Cynac said that he was going to talk to his state attorney general that the invention company stopped calling.

This story shows the importance of a healthy skepticism. I know it is an old saying, but if something seems too good to be true, it probably is.

I don't reject everything that seems too good to be true. But if something seems too good to be true, like that big piece of cheese that just showed up one morning, I am a little more careful.

Facilitated Communication

Autism is a profound developmental disability that involves an impaired ability to relate to other people. In addition, autistic children often suffer from language difficulties and often rigidly adhere to routines.

During the 1980's, an interesting discovery came out of Australia. When adults held the fingers of mentally disabled autistic children over keyboards, messages were typed out that were far beyond the abilities the children were thought to possess. Children as young as 6 or 7, who were thought to have little or no writing or reading skills, were able to type messages that were more typical of children who were much older. Almost everyone who saw this occur was stunned because it appeared that not only were these children not mentally disabled, but they seemed to be extraordinarily gifted at communication.

This was indeed a promising development for the autistic community. It now appeared that these children had a handicap that was simply an inability to express their thoughts. Once they were assisted by an adult at a keyboard, they were able to communicate. A wise person would not dismiss this new discovery as impossible, but would approach it with a healthy skepticism.

Wouldn't it be a smart move to check and see if the adult who was holding the child's fingers was doing the communicating?

Back in the United States, a small number of educators heard about this new technique and tried it on the autistic children they were working with. The educators held the fingers of the children over keyboards, and just as in Australia, the American children appeared to type messages that shocked their teachers. Children who were previously thought to possess no reading or writing skills, were not only typing complete sentences, but were also expressing thoughts that required advanced reasoning ability.

As word of this miracle spread across the country, thousands of autistic children were provided with keyboards and adult facilitators. In addition, education plans were dramatically changed because the typing that appeared to originate from the children, made it clear that they were functioning at a higher intellectual level.

Before the schools spent millions of dollars, wouldn't it have been smart to see if the children were really the ones who were doing the typing?

This new technique was called facilitated communication and it spread rapidly across the United States. Unfortunately, this occurred before any scientific tests were done to see if it really was the children who were communicating through the keyboard. Then something happened that made it imperative that these tests be done. Messages started appearing on computer screens that accused parents and others of abuse.

Fathers were put in jail and children were taken away from their parents based on messages that were typed on the keyboard. As the accusations of abuse entered the judicial system, the courts demanded that the question of who was communicating be determined scientifically.

What is so sad about this story is that it was so easy to scientifically test facilitated communication. All that needed to be done was to show the child and the adult different pictures and then see what was typed on the keyboard.

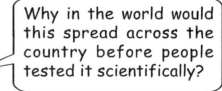

Why in the world would this spread across the country before people tested it scientifically?

When the scientific tests were completed, the results were shocking. No matter how many times pictures were shown to the adult and the child, what was typed on the keyboard was the name of the picture that was shown to the adult. These results made it clear that the adults were the ones who were authoring the communication, not the children.

What was even more shocking than the results of the scientific experiment was the reaction of many of the people who were using facilitated communication. They chose to ignore the scientific evidence and continued to believe that it was the children who were actually communicating through the assisted typing.

People often believe what they want to believe and quickly ignore what math and science tell them. What was especially disturbing about facilitated communication, was that during the typing, while the adults had their eyes glued to the keyboard, the children often were looking out a window.

The educators really should have known to test facilitated communication scientifically before they started using it. At least the courts required scientific proof and freed the falsely accused people.

Discussion questions:

1) Why did facilitated communication spread so quickly?

2) Why did many people ignore what the scientific experiments told them about who was doing the communicating?

3) Is it ethical to continue to practice facilitated communication after the scientific method showed that the children were not the ones communicating?

Herd Immunity and Vaccinations

The heartbreaking consequences of childhood diseases such as smallpox, polio, and measles are, for the most part, things of the past in the United States. Because of overwhelming public support and participation in an aggressive vaccination program in the last half of the 20th century, parents no longer live in fear of the epidemics that brought misery and death to their children.

Because of the success of the vaccination program in the United States, the devastating effects of childhood viral diseases are fading from memory. No longer do we see children walking in braces, or wards full of frightened children inside iron lungs because polio robbed them of the use of their bodies. The rib-cracking effects of whopping cough, although not eliminated, are now rare occurrences. In addition, measles no longer brings misery to hundreds of thousands of children each year.

Because these memories are fading from view, and because there are rare side effects of live vaccines, an alarming number of people are currently not vaccinating their children. More and more people are now becoming convinced, despite scientific evidence to the contrary, that vaccinations are to blame for a myriad of health problems including seizures and autism. Because of this belief, they are choosing to not have their children vaccinated. The scientifically established side effects, although sometimes serious, are generally thought to be an acceptable price to pay to guard against the much more serious consequences of an epidemic.

While it is important to keep an open mind to the possibility that the medical establishment is wrong concerning its advice about vaccinations, it is also very important to have a healthy skepticism when it comes to claims about vaccination induced health problems.

Sometimes I focus on the claims of parents with disabled children and forget about what would happen if epidemics of childhood diseases started again.

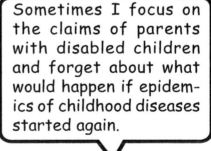

All 50 states allow parents to chose not to vaccinate their children before they enter school if there is a medical reason, such as a compromised immune system or allergies to vaccinations. In addition, close to a third of the states also allow exemptions to mandatory vaccinations for personal or philosophical reasons.

We are protected from epidemics when a large percentage of the population is vaccinated. This protection is sometimes called a "herd immunity". If a very small number of parents chose to not vaccinate their children, the "herd immunity" is still effective, but when these numbers reach a certain level, then the "herd immunity" begins to break down and others are put at risk.

There have been several serious outbreaks of diseases because of the breakdown of "herd immunity". One such occurrence happened in Southern California in 1990 when there were over 40,000 cases of measles and close to 100 deaths. There are also many areas that have a real potential for the breakdown of the "herd immunity". One such community is Ashon Island in the state of Washington. In Ashon Island, 18% of primary grade school children have not been vaccinated against many of the childhood diseases. These children and others in the community are at risk if an epidemic occurs.

When there are groups of unvaccinated children who are in close contact with each other, not only are unvaccinated children at risk, but there is also a risk to many other people in the community. Infants, the elderly with weak immune systems, fetuses, and even vaccinated children (where the vaccination unknowingly did not take), are all susceptible to infection if an epidemic occurs.

I always wondered why children were being vaccinated for rubella because I had always heard that it wasn't a serious disease. Then I found out that in the 1960's, over 50,000 children were born with serious birth defects such as small heads, deafness, or blindness because their mothers had rubella while they were pregnant.

I know it is a hard decision to make when the health of a child is at stake, especially when there are known side effects to vaccines. But because the media has broadcast several stories that imply a connection between vaccinations and autism, seizures, and unexplained deaths, the public has increasingly turned against vaccinations. Because they are frightened, they don't make their decision in a scientific manner.

Remember to maintain a healthy skepticism when these kinds of claims are made. Keep an open mind, but remember that the consequences of not vaccinating against childhood diseases are so serious that the decision must be made in a scientific, not an emotional manner.

Discussion questions:

1) Explain the meaning of "herd immunity".

2) Unvaccinated children can put others at risk. Who is put at risk and why?

3) When scientists are consulted about whether children should be vaccinated, they usually say that "we must look at the big picture". What do they mean by saying "We must look at the big picture"?

The Simple Lie-Detector Test
(Applied Kinesiology)

Some people believe there is a very simple method of testing whether a person is telling the truth or lying. The person who is being tested is told to hold his arm out to the side. The questioner starts pressing down on the arm as the individual answers a question. If the arm goes down easily, then the individual is lying. If the arm is difficult to push down, then the person is telling the truth.

Some people even use this test to find out whether a substance is poisonous. They claim that when a "bad" substance is placed in the hand, the negative vibrations weaken the arm. On the other hand, when a good substance is placed in the hand, the arm is strong.

Applied kinesiology, of course, is total nonsense. It has been shown scientifically that the strength of the arm is entirely dependent on what the tester expects to happen.

It is important to maintain a healthy skepticism. As you can well imagine, there can be very serious consequences if someone is relying on the arm test to determine what is poisonous.

Science has answered the question as to whether applied kinesiology is legitimate or not. When the tester does not know what is in the hand of the person whose arm is held out, even very poisonous substances are identified as safe. In addition, harmless items are identified as poisonous when the tester does not know what is in the hand. Applied kinesiology has been thoroughly tested and has been found to be ineffective.

Discussion questions:

1) Why is belief in applied kinesiology not a harmless belief?

2) How would you set up a scientific experiment to test applied kinesiology?

3) What would Occam's Razor tell you about applied kinesiology?

You Must Maintain a Healthy Skepticism
Level 1

1) N-Rays were said to be a kind of radiation that was given off by almost every substance. They were discovered by Rene' Blondlot around the time x-rays were discovered. Soon after Blondlot announced his discovery, a large number of scientists confirmed that N-rays did in fact exist and had many interesting properties. Are N-rays real?

2) Some people used a method of predicting the future called augury. This practice required that a certain kind of animal be torn apart and its entrails examined. A prediction was made based on the appearance of the entrails. What animal was used in this method of predicting the future? Is it still being used today?

3) The developers of the "radio detection and ranging" system claim that their machine can not only detect objects beyond the range of vision, but can also tell how far away they are, their size and shape, their speed and direction of travel. Can this system really do all that the developers claim it can do?

4) Power pyramids are said to be able to do incredible things such as keep razor blades sharp, prolong life, and make polluted well water safe to drink. Do pyramids have any kind of special powers?

5) A company that makes a special bracelet claims that its bracelet can restore health, relieve cancer pain, and provide an array of other health benefits by removing the buildup of positive ions in one's body. Can this bracelet do what the company claims it can do?

You Must Maintain a Healthy Skepticism
Level 2

1) Reflexology is the art of diagnosing the health of an individual by examining the bottom of his or her feet. Practitioners claim that different parts of the bottom of the foot correspond to different organs in the body. Is there any truth to these claims?

2) Levitation involves floating in the air in apparent defiance of the laws of gravity. Many magicians use the illusion of levitation as part of their act, but several individuals claim they have the ability to actually defy the laws of gravity. Do some people have the power to really levitate?

3) Phrenology is a medical testing technique where the size and location of bumps on the head are used to determine a person's personality. Is phrenology legitimate?

4) A company advertises that it is selling a device that eliminates the need to pay monthly charges for cable or satellite signals. The device, they say, is perfectly legal, and only cost $19.95. The company says that the device is able to pull signals directly from the air. Is this company involved in a scam, or are they telling the truth?

5) Radionics is the study of and use of vibration waves to detect and cure diseases. Thousands of people around the world use various types of radionics machines to detect and alter negative wave frequencies that they claim are responsible for many illnesses and deaths. Proponents claim that radionics machines can remove bacteria, poisons, heavy metals, pollution, and various other negative influences from the environment. Can radionics machines do what their proponents say they can do?

You Must Maintain a Healthy Skepticism
Einstein Level

1) Oxygenated water or "aerobic" oxygen is a fairly new product that many believe can cure diseases by increasing the amount of oxygen that reaches the cells. The believers in oxygenated water point to the decreasing oxygen content of the earth's atmosphere and the low levels of oxygen in junk food as reasons why people should take extra oxygen. Is there any truth to the claims that oxygenated water can improve one's health?

2) Believers in the powers of acetylsalicylic acid claim that it is a miracle substance. They say that it not only helps prevent heart attacks, but will also help relieve the pain and swelling of arthritis. Believers in the power of acetylsalicylic acid also claim that it will lower fevers and may prevent some strokes. Can acetylsalicylic acid really do what its proponents say it can do?

3) Iridology is a method of determining the health of each part of the body by looking at the eye. The iris is charted and each part is said to correspond with specific parts of the body. If there are spots or marks on certain parts of the iris, it means the corresponding part of the body is currently diseased or will become diseased. Is iridology a legitimate science?

4) Many people are enchanted with crystals and claim that they can not only heal illnesses, but are also a key to infinite wisdom. Some even believe that crystals give the holder precognitive abilities. Do crystals hold any special powers?

5) Perpetual motion machines are devises that provide free energy once the machine is put in motion. Hundreds of inventors have pursued this dream because such a machine would provide all the energy the world needs at little or no cost. Has anyone ever made a working perpetual motion machine?

Chapter 7

Don't be Fooled by Statistics

The manipulation, deception and outright lying that can accompany the use of statistics makes it imperative that children learn how to interpret them. "Statistical thinking will one day be as necessary for efficient citizenship as the ability to read and write." -H.G.Wells-

I made both graphs on my computer, and then I made one that I thought was a fair graph. I used 6 hours as a range on my fair graph because that is the amount of time students have between school and when they need to a sleep.

This is an honest graph. It looks like the teacher's graph was a little bit closer to the truth than the student's graph.

Homework Each Week

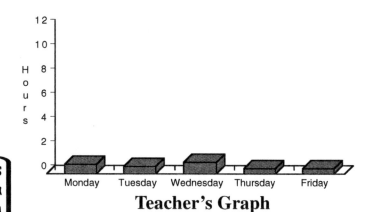

Student's Graph

Homework Each Week

Teacher's Graph

Homework Each Week

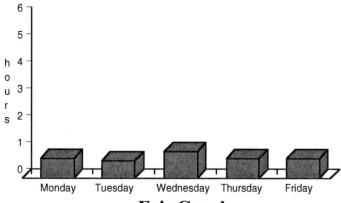

Fair Graph

Painting Bull's-eyes Around Arrows

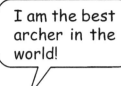

By all appearances, this archer is an excellent shot. The mouse was right to be suspicious about how far away the archer was when he shot the arrows, but even the fact that the archer was 50 yards away doesn't mean that his skills are extraordinary. The truth is not apparent until you see what happened ten minutes earlier.

Ten Minutes Earlier

The story of the cheating archer illustrates a way people use statistics to fool people. This technique is called "painting bull's-eyes around arrows." In the next fictional story, notice how leaders in a small town try to convince people that it is just as cheap to fly out of their small town as it is to fly out of Chicago. The leaders paint bull's-eyes around arrows.

(Fictional Story)

An airport in a small city was tired of losing customers to Chicago airports. Most of the people in this small city would fly out of Chicago because they believed that flights were cheaper there. If a city airport loses customers to other cities, it loses money and can be forced to shut down, so the city leaders were very concerned about the loss of customers.

We need to get more people to fly out of Smallville. We need to show them statistics that will make them believe that our tickets are the same or cheaper than Chicago's.

I have the perfect solution. We'll be truthful, but we'll manipulate the statistics.

Fly Smallville and Save Money

Smallville to Boston---------------------------$250	
Chicago to Boston--------------------------$278	
Smallville to Los Angeles------------------$312	
Chicago to Los Angeles--------------------$480	
Smallville to Orlando----------------------$295	
Chicago to Orlando------------------------$305	

The city of Smallville started an advertising campaign to show people that they didn't need to drive all the way to Chicago to get reasonably priced plane tickets. It placed advertisements in the local newspaper comparing ticket prices from Smallville and Chicago to three other cities.

It looks like it would be foolish to drive all the way to Chicago to get tickets. The prices from Smallville are about the same or cheaper than those from Chicago.

What Smallville doesn't tell you is that it is almost always cheaper to fly out of Chicago. Smallville went through a list of 500 flights and picked out the only ones where it was cheaper to fly from Smallville. This is a classic case of painting bull's-eyes around arrows. They aren't really lying, but they are being deceptive.

Discussion Questions:

1) Do you think it is ethical to do what Smallville did?

2) How long do you think the advertising campaign will fool people?

3) Make up another example of how "painting bull's-eyes around arrows" might be used to distort the truth.

The Doctors Who were Fooled

Because it is very easy to be fooled by statistics, millions of people are misled each day by information that they read. This can have serious consequences, but it is especially dangerous when doctors are misled by statistics.

When a new drug is tested, people must look at the results and decide whether the drug is effective. A group of scientists decided that it would be interesting to test doctors to see if they could tell how effective a drug was by looking at statistics on that drug.

Let's give 100 doctors a test to see if they know how to tell if this heart drug is effective at preventing heart attacks.

We'd be happy to take that test. We want to make sure we know how to interpret statistics.

The scientists told 50 of the doctors a story about a heart drug where .3% of the patients getting the drug had fatal heart attacks, while .4% of the patients who didn't get the drug had fatal heart attacks.

There was only a .1% difference between the number of deaths in the two groups. That is only a 1/10 of 1% difference. That drug sure didn't help much.

The scientists then gave the other 50 doctors the same information about the heart drug, but they presented it in a slightly different way. They told the doctors that the heart drug caused a 25% decrease in deaths.

This drug is tremendous! It caused a 25% drop in deaths.

Even though it doesn't appear to be the case, the information that the mathematicians gave each group of doctors was identical. A drop from .4% to .3% was the same as a 25% drop. Even doctors can be fooled.

After the scientists looked at what each group of doctors said, they found that many doctors were fooled. The doctors who were given the information of a 25% drop in deaths were much more likely to say they would prescribe the drug. This experiment showed that the way numbers are presented can change what people believe--even highly educated doctors!

Discussion Questions:

1) Explain why the doctors were fooled.

2) Why is it dangerous when doctors get confused about the results of drug tests?

3) Make up a story using numbers and percents that might confuse people.

Making the Crime Rate Look Better
(While it gets worse)

(Fictional Story)

A city was frustrated by its high murder rate, so the city council hired a new police chief to see if better leadership would have a positive impact on the murder rate. The number of murders that had occurred in the previous four years are shown below:

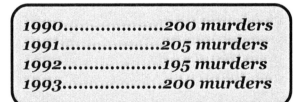

> *1990....................200 murders*
> *1991....................205 murders*
> *1992....................195 murders*
> *1993....................200 murders*

The new police chief was hired at the end of 1993 and he agreed to return four years later to show the city council that he was having a positive impact on the crime rate.

At the end of 1997, the police chief was gathering statistics so he could report his progress to the city council. The council was grumbling because the local newspaper was reporting that there appeared to be more criminal activity in the city. Unfortunately for the police chief, the number of murders had significantly increased. The year after he arrived, murders went from 200 to 300. The next year, they jumped to 375 and the third year they increased to 425. In the fourth year they reached the 450 mark.

1994....................300 murders
1995....................375 murders
1996....................425 murders
1997....................450 murders

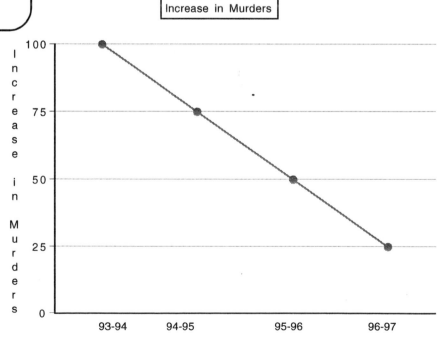

The city council was impressed with the chart and congratulated the chief on the impressive decline in the yearly increase of murders. They asked to be updated four years later. At the end of 2002, the police chief again had a problem. The number of murders continued to go up, and this time the amount of increase was even climbing.

1998.....................550 murders	
1999.....................655 murders	
2000.....................765 murders	
2001.....................880 murders	

This time even the amount of murders increased each year! There must be some way of looking at this in a positive way.

1998..................550 murders
1999..................655 murders
2000..................765 murders
2001..................880 murders

I present to you today some strong evidence that the percent of increase in murders is steadily declining.

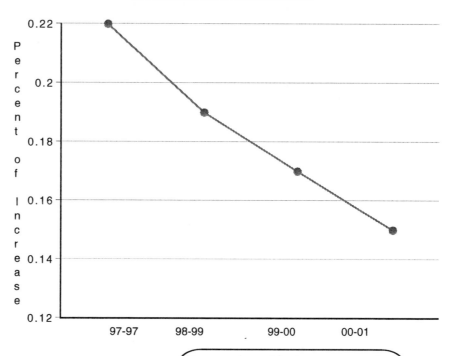

Percent of Increase of Murders

Remember that you can determine the percent of increase by making a fraction. The amount of increase is the numerator and the original is the denominator.

The police chief didn't lie to the city council, but he gave such a distorted picture of the truth that it wasn't much different than lying. This story is a dramatic example of how statistics can be manipulated to show almost anything you want.

The message of this story isn't that you shouldn't pay any attention to statistics. The story of Dr. Snow and cholera shows how important statistics are. The lesson to be learned is that you must analyze statistics very carefully.

Discussion Questions:

1) The police chief said he was only trying to put the best face on the situation and that he didn't lie. Was the police chief acting in an ethical manner?

2) What questions should the city council have asked the police chief?

3) Should the police chief have been proud that the increase in murders dropped each year in his first four years? Why or why not?

Which Car is Better?

One way to measure how well a company makes cars is to see what percent of the cars are still being driven after 10 years. *(Story is based on actual advertising in the early 1980's)*

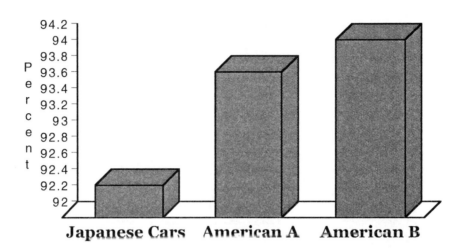

Cars on the Road After 10 Years

When I look at this graph, I think that the quality of American cars far exceeds the quality of Japanese cars.

Make sure you look at the scale when you look at bar graphs. In this graph, all three car manufacturers have pretty much the same percentage of cars on the road after ten years, but the graph makes it appear as if the quality of the American cars far exceeds the quality of the Japanese cars. If I were a Japanese car manufacturer, I would show you a graph like the one on the next page.

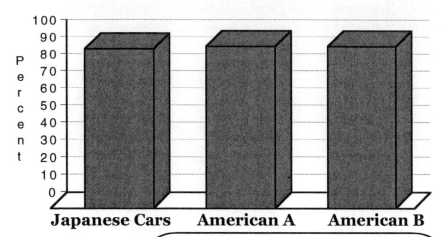

Cars on the Road After 10 Years

Now it is clear that there is very little difference between the car companies. Remember that bar graphs are visual and they can be easily misinterpreted if one does not look at the scale and intervals. Both graphs use the same information, but that information is presented differently.

Discussion Questions:

1) Should this kind of deceptive advertising be allowed?

2) If a Japanese car manufacturer had only 70% of its cars on the road after 10 years, while American manufacturers had 95%, how could a bar graph be drawn that would make the differences seem very small?

3) Bar graphs are visual and therefore they can deceive. Why is it so easy to fool people using bar graphs?

Don't be Fooled by Statistics
Level 1

1) Ariel has a chance to work at Company A or Company B. Company A pays its employees an average salary of $32,000, while Company B pays its workers an average salary of $25,000. If Ariel cares only about salary, give her advice about which company to pick. Are there any questions she must ask about the average salaries at each company?

2) Heidi wanted to be the new police chief in Chicago. During her interview with the mayor, she mentioned that during the five years she was police chief at another city, only 32 people were murdered. Because hundreds of murders were occurring in Chicago each year, Heidi said that she was confident her experience at keeping the murder rate low would help improve the murder rate in Chicago. If you were the mayor of Chicago, what kinds of questions would be important to ask Heidi to be sure that she had the ability to lower the murder rate in Chicago?

Use the following information for problems 3-5:

The citizens of Dubuque, Iowa, just heard about a rate increase for cable television. They are furious because the increase in 2003 is higher than it has ever been. The city council wants information about the increase.

Basic Cable Rates:

2000: $31.55
2001: $33.13
2002: $36.95
2003: $41.95

3) Pretend that you are employed by the cable company. Present a bar graph that makes it appear that the increase is reasonable.

4) Pretend you are part of a consumer group. Present a bar graph that makes it appear that the increase is very large.

5) Pretend that you work for a newspaper and want to present a fair bar graph that shows the rate increase.

Don't be Fooled by Statistics
Level 2

1) Scott was making a $50,000 per year salary in 1995. Because the company was doing poorly, they cut his salary by 50% in 1996. In 1997, the company told Scott that because they cut his salary by 50% in 1996, they decided to restore it by giving him a 50% raise. Will Scott's salary be restored to its original level? Why or why not?

2) A school board member received a report on student achievement as shown here. Why is this graph somewhat deceptive?

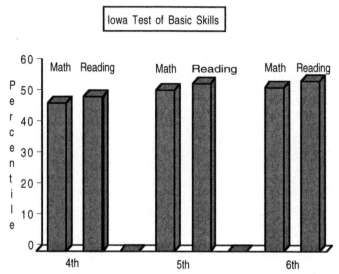

3) Use the information in question #2 and draw a graph that more fairly represents student achievement.

4) Does this graph show that Larry makes significantly more money each year than Ted?

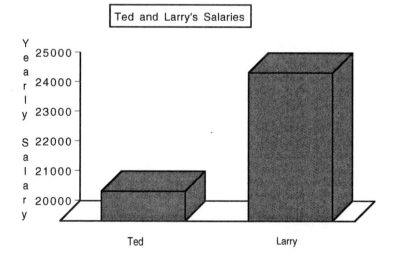

5) Seth claims that when he visited his favorite city during the summer of 1998, the crime rate dropped because he sent out good vibrations. His proof is that in the summer of 1997 there were 92 robberies and in the summer of 1998 there were only 80 robberies. In addition, Seth noted that claims of police brutality dropped from 10 in the summer of 1997 to 6 in the summer of 1998. What kind of statistical lying is Seth doing?

Don't be Fooled by Statistics
Einstein Level

Use the following information for questions 1-3:

> *Susan is president of an internet company. As part of her job, she must write a report each year telling how much money the company made. In 1998, the year before she was president, the company made 25 million dollars. This year, while she was president, the company made 50 million dollars.*

1) Susan decided to draw circles to represent profits. She wanted to be very fair when she drew her circle graphs, so she drew 1998 with a radius of one inch. Because profits doubled in 1999, she drew that graph with a radius of two inches. Does Susan's graph honestly show the change in profits from 1998 to 1999? Why or why not?

2) Susan was also thinking of making a triangle to show the increase in profits. Because profits doubled, she made one triangle with a one inch base and a one inch height and another with a two inch base and a two inch height. Is this a fair way to represent the increase in profits? Why or why not?

3) Susan had a new program on her computer that allowed her to make graphs that used spheres. She made a sphere with a 2 inch radius for 1998 and because profits doubled, she made a sphere with a four inch radius for 1999. Is this a fair way to represent profits? Why or why not?

4) A local grocery store uses the following information to prove that its prices are lower than the Giant Food Store that is ten miles away. Is this pretty strong proof that the local store's prices are lower?

	Local	*Giant Store*
Milk:	*$1.75/gallon*	*$2.00/gallon*
Cheese:	*$.20/pound*	*$.30/pound*
Bananas:	*$.29/pound*	*$.49/pound*
Apples:	*$.22/each*	*$.30/each*

5) Janelle scored at the 90th percentile on the math section of the Iowa Test of Basic Skills. Because her brother scored at the 45th percentile, Janelle is telling everyone that she is twice as smart as her brother. Why is Janelle's thinking flawed?

Chapter 8

You Must Know the Difference Between Cause and Correlation

The failure to understand the difference between cause and correlation has led to an inability to differentiate what is true and what is not true in medicine, education and in many areas of our everyday lives.

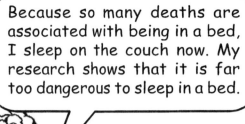

Because so many deaths are associated with being in a bed, I sleep on the couch now. My research shows that it is far too dangerous to sleep in a bed.

That researcher is scaring me.

This researcher, of course, has drawn a silly conclusion from his data. He found that a high percentage of people who die are in a bed when they die, so he concluded that beds must be dangerous. His mistake, confusing cause with correlation, is a very common mistake. The researcher's thinking error is very obvious here, but in many situations it is very easy to unknowingly fall into the trap of assigning cause improperly.

The Bendectin Nightmare

In 1956, Bendectin was approved by the FDA to treat morning sickness in pregnant women. Over the next 20 years, Bendectin helped millions of women by preventing the nausea that often accompanies pregnancy.

Late in 1979, the women who used Bendectin received some disturbing news. The National Enquirer published a frightening story that told of thousands of women who had taken Bendectin during pregnancy and had then given birth to babies with birth defects.

Merrell, the company that made Bendectin, sought to reassure those who had taken the drug by pointing out that there was extensive research showing no connection between birth defects and Bendectin use. This information was not reassuring, and the anxiety and concern of women who took Bendectin increased over the ensuing months. More and more parents of children with birth defects came forward and publicly stated their belief that Bendectin was the cause of their children's birth defects.

Because of the new safety concerns for Bendectin, it was very important to determine whether the drug was dangerous. Unfortunately, because the story in the National Enquirer brought back memories of the thousands of babies who were harmed when their mothers took thalidomide during pregnancy, there was a strong possibility that the drug would be judged emotionally and not scientifically.

It wasn't long before lawyers started filing lawsuits on behalf of women who took Bendectin. In courtrooms across the country, handicapped children tugged at the heartstrings of jurors and made dispassionate decision-making nearly impossible. Some juries found that Bendectin was the cause of birth defects, while others found the drug blameless.

In June of 1984, close to 1000 cases were combined into one large proceeding in Cincinnati, Ohio. It was hoped that this trial would be the final word on whether Bendectin caused birth defects. By the time the trial took place, there was still no evidence that Bendectin caused birth defects, and even though Merrill had won most of the previous trials, the drug company offered to settle the case for 120 million dollars.

Merrill was still adamant that Bendectin did not cause birth defects, but it was worried about what a jury might do after seeing hundreds of children with severe birth defects during a lengthy trial.

The plaintiffs and their lawyers decided that the 120 million dollars was not enough money, so the decision was made to go ahead with the trial. Before it began, the presiding judge set ground rules that horrified the lawyers for the plaintiffs. He made it clear that the trial would revolve around whether Bendectin caused birth defects, and that the decision would be based on science and not sympathy. To that end, he announced that the trial would be in three stages. The first stage of the trial would answer whether Bendectin caused birth defects, and only if it was decided that it did, would the trial move on to the next two stages. If it was determined that Bendectin did cause birth defects, the jury would then decide which children were harmed, and finally, what compensation was due to those children.

To ensure that the first part of the trial was fair, the presiding judge banned the use of the word thalidomide, and also barred children with visible handicaps from the courtroom.

This trial will use science to determine whether Bendectin causes birth defects. The jury will not make its decision based on emotion. Because this trial is about Bendectin and not thalidomide, any mention of the word thalidomide is forbidden. Also, because children who have noticeable birth defects might encourage a sympathy vote, they are excluded from the courtroom during this phase of the trial.

After a long trial, the jury took less than 5 hours to find that Bendectin did not cause birth defects. Science, not emotion, was used to determine the truth.

The decision was appealed, but the jury's verdict was upheld. Although the Cincinnati case was not the final word on Bendectin, it was a powerful influence on future cases.

I don't understand something. If these 1000 women all took Bendectin, and all had children with birth defects, why did the jury say that Bendectin wasn't the cause?

You, like many people, are a victim of flawed thinking. You have to understand that each year there are about 100,000 children who are born with some kind of birth defect. When you realize that over 30 million women have taken Bendectin, it is not hard to see that thousands of children with birth defects will be born to women who took Bendectin, even if the drug is totally harmless.

What if I found 1000 women who ate oatmeal during pregnancy, who also had children who were born with birth defects. Could I then conclude that oatmeal caused the birth defects? Of course I couldn't! It is important not to confuse cause with correlation.

The real tragedy here is that pharmaceutical companies will be very unlikely to provide drugs, or do research on drugs, that might help women who suffer from "morning sickness." Merrell pulled Bendectin from the market, not because it was dangerous, but because it was too expensive for them to defend. The irony of this situation is that untreated nausea can hurt both the mother and the developing fetus.

Discussion Questions:

1) Even if Bendectin didn't cause the birth defects, the drug company does have a lot of money. Is it right for juries to award some money to the defendants because they feel sorry for them?

2) The trial judge in the Ohio case banned children with visible handicaps from the courtroom during the first phase of the trial. Why did he ban these children?

3) Was it ethical to ban children with visible handicaps from the courtroom?

Parent-Teacher Conferences

Research has recently established what many teachers have long known. Children whose parents attend parent-teacher conferences do better in school. The results of this research make sense because children usually are more successful students when parents are involved in the educational process.

This kind of research is what I call "duh" research. The results are so obvious that I wonder why money was spent studying the issue.

Sometimes even obvious things need to be studied, just to make sure. In addition, the research might tell us that attending conferences is very helpful or just slightly helpful.

After the research was publicized, a few schools wanted to use the information to help their students raise their level of achievement, so they started discussing what the research meant.

How can we use this information to help children?

The research shows that the children of parents who attend conferences do better academically.

I know! We can pay parents to come to conferences. That way we will be sure to have a high percentage of parents who attend.

This school made a very common mistake when they decided that paying parents to come to conferences would increase student achievement. Their mistake wasn't that it is wrong to pay parents to come to conferences, (although it probably is wrong). What the school did wrong was to confuse cause and correlation.

It is very unlikely that this school's actions will help children because there is something special about parents who come to conferences. They care enough about the education of their children to come to school and discuss their child's academic progress. They value education! On the other hand, parents who don't come to conferences, or must be paid to attend, probably do not have education as a priority in their homes.

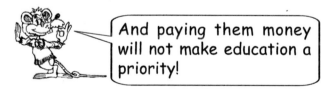

Discussion Questions:

1) The children of parents who come to parent-teacher conferences do better in school. Therefore, to increase achievement, our goal should be to have 100% attendance at conferences. How does this statement confuse cause and correlation?

2) If a survey showed that the highest achieving students in the school all had expensive backpacks, does that mean that expensive backpacks encourage better achievement? Why do you think children with expensive backpacks might do better academically?

3) Sasha noticed that people who play basketball are tall. She concluded that if you play basketball, you grow more than those who don't play basketball. What is wrong with Sasha's logic? How is she confusing cause and correlation?

The Bermuda Triangle

The legend of the Bermuda Triangle has led to the belief that a triangular shaped area in the Atlantic Ocean is very dangerous for ships and planes to travel through. Many people believe that something strange is going on in the Bermuda Triangle because it has been reported that a large number of planes and ships have mysteriously disappeared or sunk in this area.

Some of the bizarre explanations for the "mysterious events" in the Bermuda Triangle involve extraterrestrials, strange magnetic fields, and antigravity fields. The more rational explanations include hurricanes, pirates, storms, and human error.

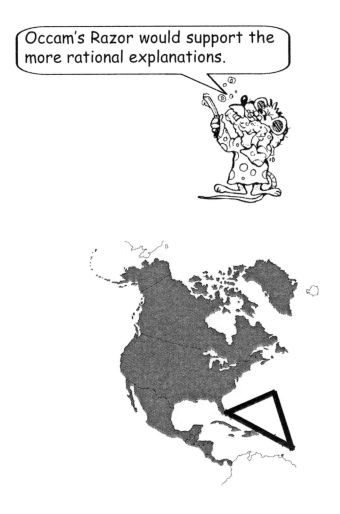

Some people who believe in the dangers of the Bermuda Triangle accept the rational explanations for the shipwrecks and disappearances, but say that something strange must be going on because of the large number of problems that have occurred in this area.

When I look at some other triangles in the ocean, they don't have nearly the number of ships and planes that have either crashed or disappeared. Doesn't this mean that the Bermuda Triangle is dangerous?

You are confusing cause and correlation. There is a lot of ship and plane traffic in the Bermuda Triangle, so naturally you would expect that there would be more plane crashes and shipwrecks there. In addition, the area is known for its severe weather.

It is often said that most accidents occur in a person's home. Does that mean that it is safer to be at a friend's house?

Of course not. The reason most accidents occur in the home is because you spend most of your time there.

The legend of the Bermuda Triangle started shortly after a Navy training mission off the East Coast of the United States ended in tragedy. Four training planes, plus a lead plane piloted by Lt. Charles Taylor, vanished in a severe storm. Many explanations have arisen to explain why the five planes never returned, but there really is no need to blame mysterious forces.

The training planes were not capable of navigating on their own, so if Lt. Taylor had problems with his compass, it is very likely that all five pilots would have become disoriented and run out of gas before they could find their way back to land.

So you're saying that this simple explanation is more likely than the five planes being caught in an antigravity vortex?

Yes! Now I'm not saying that it is impossible that an antigravity vortex -- whatever that is--caused the five planes to plunge into the ocean, but I think it was just plain old gravity that caused the problem.

Actually, I made up the phrase antigravity vortex. I don't even know what it means, but it sounds mysterious and has a scientific ring to it. People believe my theories more and think I'm smart when I use fancy sounding words like that.

Discussion Questions:

1) If you drew a triangle between Boston, New York City, and Philadelphia, you would find that thousands of car crashes have occurred there. You could even call it the Northeast Triangle and start talking about how many mysterious deaths occurred there. Explain why a believer in the "Northeast Triangle" would be confusing cause and correlation.

2) There are many triangles in the ocean that are as big or bigger than the Bermuda Triangle, but have far fewer shipwrecks. Does this mean it is safer to travel through these triangles than the Bermuda Triangle?

3) An author of a book on the Bermuda Triangle wrote that more than 50 ships and 20 planes have gone down in the Bermuda Triangle in the past 100 years. He made the claim that this proves there is something mysterious going on in this area. Why is his reasoning wrong?

Music and Higher Achievement

Several experiments have shown that there may be a connection between musical training and intellectual development. One recent experiment divided preschool children into three groups. The first group received keyboard instruction and singing lessons; the second group received computer lessons; and the third group received no lessons.

The results of this experiment were very interesting. The group that received the keyboard training and singing lessons scored 34% higher on tests that measured spatial-temporal ability. This connection between music training and intellectual abilities makes sense because music has clear mathematical relationships and proportions.

The children with keyboard training and singing lessons did better at mazes, copying patterns of blocks, and making geometric shapes out of puzzles.

Another often mentioned piece of evidence for the connection between music lessons and higher intellectual development is the fact that children who have had music lessons, score significantly higher on SAT tests when compared to children who have not had music lessons.

At first glance, these test score results appear to prove that music lessons lead to better academic achievement, but we have to be careful not to confuse cause and correlation.

We need to ask ourselves whether there is something special about families that provide music lessons for their children. Because music lessons require an enormous investment in time and money, it is clear that there is something special about these families.

It may turn out that music lessons do lead to higher achievement---- I happen to think that they do. But we can't say they do, based only on the fact that children who take music lessons do better in school.

I studied the golf skills of children who ride to school in new luxury cars and found that they were much better golfers than children who ride to school in older cars. If you want to improve your golf game, convince your parents to buy a new luxury car!!

Discussion Questions:

1) Einstein says that music has clear mathematical relationships and proportions. What does this mean?

2) How is the mouse confusing cause and correlation when he talks about luxury cars and golf?

3) Do you think musical ability and mathematical ability are related?

How Tobacco Companies Tried to Confuse Cause and Correlation

Throughout the 1900's, tobacco companies have lied about the dangers of tobacco. In the middle part of the 20th century, when almost 90% of lung cancer patients were found to be smokers, many people thought that the tobacco companies would finally admit that cigarettes caused lung cancer. They did not. The tobacco companies claimed that just because there was a correlation between smoking and lung cancer didn't mean that smoking caused lung cancer.

I know the beds didn't cause the people to die in the cartoon at the beginning of the chapter. I am claiming that the correlation between smoking and lung cancer may be the same thing. Maybe people who are susceptible to lung cancer are more likely to smoke.

I'm impressed that you are keeping a straight face when you say that---that takes real talent! Oh, by the way, instead of saying the tobacco companies lied, I like to say they have prevaricated. If I say they've lied, it makes me appear rude.

When scientists tried to determine whether tobacco causes lung cancer, they asked themselves several questions.

1) **Is there a correlation?**

2) **Does the risk of lung cancer go up when the amount of smoking is increased?**

3) **Is the connection between lung cancer and smoking seen in other countries and in different groups of people?**

4) **Are different income groups effected in a similar manner?**

5) **Do animals studies show a connection?**

6) **Is it biologically plausible that smoking causes lung cancer?**

Discussion Questions:

1) Tobacco companies say they are attempting to stop teenagers from smoking by distributing pamphlets that ask teenagers not to smoke. The pamphlets contain statements such as: "Smoking is an adult activity like driving, please wait until you are an adult before you make a decision about smoking." Is this an effective way to prevent teenagers from smoking? Why?

2) What does it mean to prevaricate?

3) Tobacco companies used to say that the connection between smoking and lung cancer was only a correlation. What did they mean by saying the connection between smoking and lung cancer was only a correlation?

Cause and Correlation
Level 1

1) Samantha, who was 5 years old, noticed that a rooster crowed every morning right before the sun came up. She announced that she knew why the sun came up each morning--the crowing of the rooster made the sun rise. Explain how Samantha is confusing cause and correlation.

2) A hillside was causing problems for a city. Whenever it rained, mud and small rocks would pour down the hillside into the yards of people who lived at the bottom of the hill. An inventor claimed that he had found a way to make artificial grass that would stop this damage from occurring. The city decided to test his invention on a part of the hill. After two months, no yards below the artificial grass had any rocks or mud slide into them. The inventor declared that his product successfully passed the test and this proved his invention could prevent mud and rocks from sliding into yards. Which of the following, if true, would make you believe that the inventor was making a claim that he shouldn't have been making?

a) The people who lived below where the artificial grass was placed didn't care that much whether they had mud in their yards.
b) It didn't rain during the 2 month test.
c) The inventor often said things that people don't believe.

3) A researcher studied the brains of 300 people who were in prison for violent behavior and found that they all had clear evidence of old brain injuries. The researcher then declared that old brain injuries lead to violent behavior. Which of the following should the researcher have done before he made his declaration?

a) Test the brains of 1000 more violent people to see if they also have old brain injuries.
b) Check to see if his test subjects really were violent.
c) Look at the brains of people who aren't violent and see if their brains also have signs of old injuries.

4) Because most fatal car accidents occur less than 30 miles from home, it is safer to travel when you are more than 30 miles from your home. Why is this logic faulty?

5) True or false: If all A's are also B's, then all B's must be A's.

Cause and Correlation
Level 2

1) The cancer rates in two countries are being compared to see if diet might be connected to the cancer rate. Country A is very poor and the people there subsist mostly on grains and occasional vegetables. Country B is a very wealthy country whose people have a diet that is heavy in meats and dairy products.

Cancer was found to be very rare in Country A, while Country B had almost a fifth of its deaths caused by the disease. From this information, can you conclude that the people in Country A have a diet that helps protect the population from cancer, while the diet of Country B encourages the development of cancer? What is your guess as to why Country A has a low cancer rate?

2) Seth wanted to know the safest place to live in his community, so he placed a dot where each death occurred. Much to his surprise, he found that he lived in an area where well over 100 people die each year. Give some possible explanations for the high death rates on the map that might help put Seth's mind at ease.

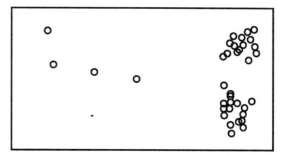

3) If it turned out that the death rate in Florida and Arizona was higher than the death rate in other states, it wouldn't mean that living in those two states caused people to be more likely to die. What would be a possible explanation for a higher death rate in Florida and Arizona?

4) The Sloan Bike Company is claiming in an advertisement that its bikes are much safer than Ketter bikes. Does the information in the ad, if true, prove that Sloan bikes are safer?

"Only 11 people were killed riding Sloan bikes last year, while 268 people died riding Ketter bikes. As a matter of fact, in every year since 1950 there have been at least twice as many deaths on Ketter bikes than on Sloan bikes. If you want a safe bike, it has been statistically proven that you should buy a Sloan bike."

5) A person decided to use a low-calorie sweetener to help lose weight. Which of the following studies should lead this person to question whether this sweetener can really help him lose weight?

a) This sweetener has been shown to cause a craving for carbohydrates.
b) This sweetener has been shown to cause cancer in mice.
c) In animal studies, 5% of the hamsters that were given the sweetener did not lose weight.

Cause and Correlation
Einstein Level

1) Several parents whose children are autistic blame the autism on vaccinations. Explain why it is possible that they are confusing cause and correlation.

2) In the 1990's, there was a worry that cell phones might cause brain tumors. This belief started because hundreds of people who had brain tumors used cell phones on the same side of their head as the location of the tumor. What important information must you have before you can say that cell phones cause brain tumors?

3) An army general is making a case for how safe it is to be in the army--even during a war. The numbers he is quoting are accurate. What is wrong with the general's logic?

The death rate for New York City in the year 2000 was 760 deaths per 100,000 people. During this year-long war, we only had 124 deaths per 100,000 soldiers. This clearly shows that it is far safer to be a soldier at war than to live in New York City.

4) Country A and Country B both have populations of 100 million people. Fires in Country A destroy 15 billion dollars worth of property each year, while fires in Country B destroy only 1/2 billion dollars worth of property each year. Can we say that Country B has better fire prevention than Country A? Why or why not?

5) School A has students whose average score is at the 80th percentile in standardized reading tests. School B has students whose average score is at the 20th percentile in standardized reading tests. The school board in this school district is outraged because of the low scores at School B, so it has decided to transfer all the teachers in School A into School B to bring up the reading scores in School B. Does this move have much chance of helping the students in School B improve their reading scores? Why or why not?

Chapter 9

Ethical Decision Making

During your life, you will likely face situations where your ethics and your ability to maintain those ethics will influence the choices that you make. The lives of others may depend on those choices.

Great news! It will only cost $11 per car to fix the gas tank. I know that adds up to about 137 million dollars over the many years we are going to build our car, but the per car cost is very inexpensive.

The projected lawsuit cost for people who are burned to death, or are burned and survive, is only 49 million dollars. We decided that it is cheaper to pay people who are burned, instead of fixing the car.

Fortunately, most companies don't act in an unethical manner like this company. The following stories will show you examples of companies that have acted unethically and the consequences of those unethical actions.

There will also be stories of companies and people who were faced with very difficult situations and chose to act in a moral and ethical manner.

The Exploding Gas Tank

In the Accreditation Board for Engineering and Technology Code of Ethics for Engineers, under the first fundamental canon, it states that engineers shall hold paramount the safety, health, and welfare of the public in the performance of their professional duties.

As American car manufacturers rushed to develop small fuel-efficient cars that were demanded by American consumers, the engineers at one company found a serious design flaw in one of their newest cars just as it was about to enter the production phase. The engineers found that the fuel system of this car had a tendency to rupture during rear-end crash tests. This led to fires and explosions that clearly would put occupants of the vehicle at risk.

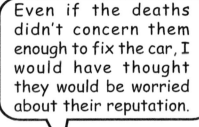

The car company's response to the newly discovered fire risk was to conduct a financial analysis to see whether they should spend the $11 per car that it would take to fix the problem. They estimated that slightly more than 2000 cars would burn and that there would be 180 deaths and 180 serious burns if the design was not changed. After assigning a value of $200,000 for each death and $67,000 for each major burn, it was determined that there would be a 49.5 million dollar benefit if the design flaw was corrected. In other words, 49.5 million dollars worth of death and injury would be prevented if they spent $11 on each car to make the fuel system less prone to catching fire and exploding.

The 49.5 million dollar figure was then compared with the known cost of 137 million dollars that they knew it would take to modify the fuel system. Using those two numbers, a decision was made that it wouldn't be profitable to modify the fuel system.

In 1978, almost eight years after the initial discovery of the design flaw, the car was finally recalled and repaired. During those eight years, hundreds of people died and thousands more were seriously burned because the car had a tendency to catch on fire when it was hit from behind.

The reason it took so long to recall and repair the cars was that the company thought it would be cheaper to leave the cars in an unsafe condition. As it turned out, when the company was sued, the juries were so outraged at the unethical decision-making that they awarded hundreds of millions of dollars to people who were burned because of the unsafe design.

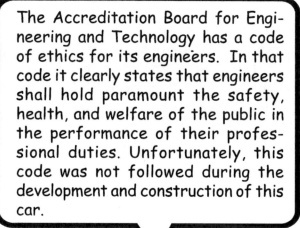

In 1978, a jury awarded 128 million dollars to a 16-year old boy who was badly burned. As it turned out, fixing the car would have been not only the ethical thing to do, but it also would have saved the company a lot of money.

The Accreditation Board for Engineering and Technology has a code of ethics for its engineers. In that code it clearly states that engineers shall hold paramount the safety, health, and welfare of the public in the performance of their professional duties. Unfortunately, this code was not followed during the development and construction of this car.

Discussion Questions:

1) When this car was built, safety was far down the list of features that attracted customers to a vehicle. How did this influence the decision to not fix the fuel system defect?

2) In this case, it was decided that the $11 cost to fix the design flaw was too high. What if the cost was $100 per vehicle, or $10,000 per vehicle? Is there ever a cost that is too high to make a vehicle safe?

3) It is possible to make a vehicle that is safe in almost any kind of crash. Why is this vehicle not being made?

Thalidomide

Thalidomide was a very popular drug in Europe during the late 1950's. It was inexpensive, appeared to be safe, and was very effective at controlling nausea. Because it was selling well in Europe, the company that manufactured the drug wanted to begin selling it in the United States. Even though every drug must be reviewed by the Food and Drug Administration before it is sold in the United States, the company expected quick approval because of its popularity in Europe.

The FDA gave the responsibility of overseeing the approval process to Dr. Frances Kelsey. Dr. Kelsey did not feel that the apparent safe use of thalidomide in Europe was proof enough to approve the drug in the United States, so she started a careful review of the research on thalidomide.

Even though you say this drug is being safely used in Europe, I need to analyze the research myself.

Don't you think that you are being overly cautious?

As the approval process dragged on, Dr. Kelsey felt more and more pressure from the manufacturer of thalidomide. They thought that the delay was not necessary and wanted the drug approved quickly. As the pressure mounted, Dr. Kelsey held her ground and made it clear that the drug would not get her stamp of approval until she was confident that it was safe.

A short while later, a disturbing report appeared in the British Medical Journal. A British Doctor was starting to see numbness and tingling in the extremities of some people who were taking thalidomide. This meant that there was the very real possibility that thalidomide was causing nerve damage.

If nerve damage was occurring, then there was the possibility that thalidomide was a teratogen. Now Dr. Kelsey had to make sure the drug was not only safe for adults, but she also had to be confident that it was not harming unborn children.

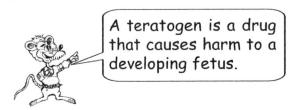

A teratogen is a drug that causes harm to a developing fetus.

Eight months later, Dr. Kelsey's worst fears were confirmed. Thousands of babies whose mothers had taken thalidomide during pregnancy were being born with very short arms and legs, or in some cases, no limbs at all.

Because Dr. Kelsey refused to recommend approval of thalidomide before she was confident it was safe, thousands of American children were saved from the appalling birth defects that were suffered by European children.

> Dr. Kelsey's courage in the face of enormous pressure from the makers of thalidomide has been an inspiration to thousands of scientists who try to do their jobs in an ethical manner.

> Even though the United State did not approve thalidomide, a small number of American children did suffer from thalidomide-induced birth defects because their mothers acquired the drug in a foreign country.

Discussion Questions:

1) Why do you think the company that made thalidomide was in such a hurry to get the drug approved?

2) Do you think families with children who were born with defects caused by thalidomide should be awarded money, even if the company that made thalidomide did not know their drug was dangerous?

3) Thalidomide is no longer being used for nausea, but it is still being used. Research the current uses of thalidomide and discuss the concerns that doctors have today.

The Unethical
Stuttering Experiment

Stuttering is a condition whose negative effects on social and emotional growth can be devastating to children. Because of the serious consequences of stuttering, the condition was thoroughly studied during the 1900's as researchers sought to find the cause and cure.

In the early 1900's, the prevailing theory was that stuttering was something that a child was born with. As more and more became known about the condition, several researchers began to question this explanation for stuttering. One of these individuals was a professor at the University of Iowa named Wendall Johnson.

Mr. Johnson, who was himself a stutterer, was convinced that stuttering was caused by the reactions of well-intentioned parents and teachers to the normal stammering of young children. When attention was drawn to word and syllable repeating, which is a normal part of language acquisition, children became so self-conscious that they could not develop normal speech patterns.

As Johnson became more and more convinced that his theory was a reasonable explanation for stuttering, he devised an experiment that could provide direct evidence that his hypothesis was true. When it came time to implement his experiment, Johnson settled on a location that the University of Iowa had already used for several research projects. It was a local orphanage that Johnson thought would be an ideal spot to conduct his research.

The first part of the study involved taking 10 stutterers and 12 normal speakers and splitting them into two groups. Half of the stutterers and half of the normal speakers were assigned to a group that was labeled the stuttering group. This group consisted of the children that the researchers would try to make stutter. The other half of each group was kept as a control group.

The group that was labeled the stutterers was interacted with in a way that the researchers hoped would induce stuttering. These children were constantly interrupted as they were speaking and warned that they were beginning to show signs of stuttering. The environment that was created for the group was designed to make the children more and more conscious and nervous about their speech.

> The children's teachers were even misled, and told that what was happening to the children was speech therapy. This was done so the teachers would participate in the experiment by constantly drawing attention to the speech patterns of the children.

The results of the experiment were dramatic. The researchers had indeed induced stuttering in five out of the six children who were talking normally when the experiment started. They also caused a deterioration in the speaking ability of three out of the five children who were already stuttering when the experiment started.

> This experiment helped clarify the causes of stuttering, but in the process, seriously harmed several children. When the researchers left, the children were left behind at the orphanage to try and cope with the social and emotional consequences that stuttering brought them. This experiment was clearly unethical and should not have been done.

Why weren't the results of the experiment published? I would think that it would be important for the world to know of this research that showed a possible cause of stuttering.

The results of the study were never disclosed because the study ended at about the time that the world first heard about the very unethical medical experiments that the Nazi's performed on living subjects. Dr. Johnson's colleagues thought that publication of this study would ruin his career.

Discussion Questions:

1) Why do you think the experiment was done at an orphanage instead of a local school?

2) This experiment gave us a lot of knowledge about stuttering. Was the experiment still unethical, even though it ended up helping thousands of children?

3) Was it unethical to not release the study after it was done?

How Ethical Decision-Making Saved a Company

In the Fall of 1982, a series of mysterious sudden deaths occurred in and around Chicago. Before long, a connection between the deaths and Extra-Strength Tylenol capsules became apparent. Someone had evidently taken Tylenol bottles from several stores, laced some of the capsules with cyanide, and then replaced the bottles on the shelves.

When unsuspecting Tylenol users took the pills, they ingested 65 milligrams of cyanide; a dose that was 10,000 times more powerful than the 6.5 micrograms that is required to cause death.

As word of the tainted Tylenol reached the public, a panic ensued. Thousands of people called hospitals after taking the painkiller and hundreds were admitted as a safety precaution. As the nation dealt with uncertainty and fear, the manufacturing firm that made Tylenol was facing a crisis that had the potential to bankrupt the company.

Johnson and Johnson had some tough decisions to make. What would be their first concern? Would it be profits or the health of their customers?

I'm glad Tylenol wasn't owned by the company that made the cars with the exploding gas tanks.

How Johnson and Johnson responded to this crisis has gone down in history as a model for corporations to follow when they are faced with a crisis. Johnson and Johnson decided to make customer safety their first priority. They alerted Extra-Strength Tylenol users across the country and recommended that they stop taking Tylenol. A few days later, Johnson and Johnson implemented a nationwide recall of 30 million bottles of Extra-Strength Tylenol. Because this recall involved over 100 million dollars worth of capsules, it cost the corporation a substantial amount of money.

Several marketing people were saying that Johnson and Johnson would never be able to sell Tylenol again. They were wrong!

Less than 2 months after the poisonings, Extra-Strength Tylenol was back in stores with triple-seal tamper resistant packaging. A month later, it had gained back a large portion of the customers who were lost because of the recall. Johnson and Johnson's actions during this tragedy also greatly enhanced its reputation as an honest and responsible corporation.

Johnson and Johnson faced a terrible dilemma when it found that several people had died after taking their product. They decided to take the morally and ethically correct approach---whatever the financial cost. As it turned out, their approach was not only morally correct, but also financially correct. Johnson and Johnson continued to prosper as a corporation because it maintained the trust of its customers.

Discussion Questions:

1) Even though Johnson and Johnson had nothing to do with the poisonings, would you hold them at all responsible for the deaths?

2) What is the difference between a microgram and a milligram?

3) How did the large expense of recalling Extra-Strength Tylenol capsules end up saving the company money?

The Building that Almost Tipped Over

William LeMessurier was an innovative and distinguished engineer whose skills were relied on when the Citicorp Tower was constructed in the 1970's. This construction was especially challenging because it had to be built around and over a church that was located on the corner of the building site where the Citicorp Tower was to be located.

LeMessurier's innovative construction techniques involved diagonal bracing, which while allowing a tremendous weight saving in construction materials, also meant that the building had a tendency to sway in the wind. Because of this, a movable 400-ton block was placed at the top of the building to counteract the expected swaying.

The light-weight design also meant that the structural integrity of the building depended on strong connections where the braces were joined. To this end, LeMessurier recommended that the braces be connected with full-penetration welds. Because this kind of welding was time-consuming and expensive, the New York contractors decided to connect the braces with bolts. Although bolts are not as strong as full-penetration welds, it was determined that they would provide more than sufficient strength to properly connect the braces.

Close to a year later, when LeMessurier found out about the switch from welds to bolts, he was not overly concerned. A few weeks later, after answering an unrelated structural question about the Citicorp Building from an architectural professor, LeMessurier decided that he would present the Citicorp design to a structural engineering class he taught at Harvard. As part of his preparation for the class, LeMessurier studied the effects of various kinds of winds on his building. Because he now knew that bolts were substituted for welds, his new study assumed that bolts were the fastening mechanism.

The results of LeMessurier's new study were alarming! He discovered that the Citicorp Building was vulnerable to a total structural failure in winds that occurred in that area every 16 years.

LeMessurier was now very worried. Not only was there a very real potential for enormous death and destruction, but his reputation and career as a structural engineer were at stake. Fortunately, making the Citicorp Building structurally sound was not a particularly difficult problem. What was difficult was summoning the strength of character to take responsibility for the problem and then notifying all concerned parties.

LeMessurier's honesty and integrity leading up to and following the discovery of structural problems in the Citicorp Building are now looked upon as a model of ethical behavior. He contacted his insurance company, the owners of the Citicorp Building, city officials and finally, the press. (Fortunately for LeMessurier, there was a city-wide press strike at the time.)

As repair work started, the drama of the situation increased when a hurricane was detected moving on a path that would bring it near New York City. Fortunately, the hurricane changed course and the repairs were completed.

LeMessurier took a responsible and ethical course of action when he learned of the structural deficiencies of the Citicorp Building. He thought that his insurance company would raise his rates, but because of the way he handled the crisis, his rates were lowered. The insurance company realized that LeMessurier prevented a disaster that not only would have killed scores of people, but also would have cost their company a substantial amount of money in claims.

Discussion Questions:

1) The construction mistake cost LeMessurier's insurance company money. Why did many people think LeMessurier's actions saved the insurance company money?

2) Give some realistic examples of unethical responses to LeMessurier's situation?

3) What is a 16-year wind?

The Radium Girls

Radium is an element with fascinating properties that was discovered by Marie and Pierre Curie in 1898. Although its lethal properties would soon become apparent, for a short time it was considered a wonder drug that could cure almost any ailment. In addition, because it would glow in the dark, it was also used to paint clock dials, doorknobs, and many other objects.

During the early 1900's, several industries prospered because they took advantage of the luminous qualities of radium. The U.S. Radium Corporation was one such company. It hired young women to paint a mixture of radium, water, and glue onto clock numbers, military aircraft controls, and other instruments. The women would use sharp, pointed paintbrushes to do their work, but after a short time the brushes would lose their point. The lack of a sharp point made accurate painting impossible, so the women were instructed to roll the tips of the brushes on their tongues before dipping into the radium mixture again.

The women were not concerned about putting radium contaminated brushes in their mouths because they didn't know that the radium mixture was dangerous. They did think it was odd that when they blew their noses their handkerchiefs would glow in the dark, but they showed so little concern for the dangers of radium that they would paint their fingernails and teeth to make them glow in the dark.

The young women didn't know how dangerous radium was, but the government, the U.S. Radium Corporation and the scientists who worked with the element were all well aware of its health risks.

In 1917, Grace Fryer started a job at U.S. Radium that involved working with the radium mixture. Fryer, like the other girls, had no idea how dangerous her new job was. She used her lips and tongue to make a sharp point on her paintbrush and joined in the fun when others painted their teeth and fingernails.

Grace quit her job in 1920 and two years later she began to experience serious health problems. Her teeth began to fall out and x-rays showed that her bones were decaying. In 1925, a doctor finally drew a connection between her health problems and her work at U.S. Radium. Around this time, a specialist from Columbia University named Fred Flynn, heard about Grace Fryer's health problems and requested permission to examine her. Shortly thereafter, Mr. Flynn and a consultant conducted a medical exam of Grace Fryer and announced that she was in excellent health.

As it turned out, Fred Flynn and his colleague worked for The U.S. Radium Corporation. They were both part of a deliberate plan started by U.S. Radium to stop the emergence of the truth about the harm that was done to women who worked at the plant.

As the years went by, it became more and more apparent that hundreds of women were either dying or becoming seriously ill because of the work conditions at many of the factories that were using radium. Eventually, Grace Fryer and four other women sued U.S. Radium. This lawsuit exposed the outrageous conduct practiced by U.S. Radium and made public their total disregard for the health of the women who worked there.

Shortly before the trial was to begin, it was discovered that several years earlier, U.S. Radium had asked a Harvard physiology professor named Cecil Driver to take a look at the working conditions at their facilities. What he found was shocking. Many workers had advanced radium necrosis, and almost everyone had unusual blood conditions. After finishing his study, Driver wrote a report that laid out changes that would help protect workers from the dangers of radium contamination.

Unfortunately, the president of U.S. Radium decided not to implement Driver's suggestions. To make matters worse, he also refused to allow Driver to publish his important findings and even threatened legal action if he did.

Another disturbing aspect to Driver's report was that it was apparent that the disregard for the health of the young women who painted with the radium mixture did not carry over to other parts of the factory. The people who were more knowledgable of the dangers of radium were usually provided with lead screens, masks, and tongs. This was further evidence that the young women were viewed as expendable.

The important thing here is that people must not know that radium has caused health problems for our workers. Can you imagine the financial harm that would come to our company if people found out?

Days before the trial was to start, U.S. Radium reached an out of court settlement by agreeing to pay each girl a lump sum of $10,000 and an additional $600 each year that they survived. U.S. Radium also agreed to pay all future medical expenses. All five women died within the next ten years.

Another tragic example of radium poisoning is the case of Peg Looney. She painted glow-in-the-dark dials for a company called Radium Dial during the 1920's. She stopped working at the age of 24 because of severe health problems. Eight days after she left work, she died. An autopsy was quickly conducted by a doctor from Radium Dial who listed the cause of death as diphtheria.

Peg Looney's parents didn't believe that diphtheria caused their daughter's death and were distrustful of Radium Dial's explanation. The real cause of death, and Radium Dial's culpability in that death, was not exposed until 50 years later.

In 1978, Peg Looney's body was exhumed and found to contain close to 20,000 microcuries of radium. This was over 1000 times the level that is considered safe by scientists.

Discussion Questions:

1) It was often said that the young women who worked with radium were expendable. What does it mean to be expendable?

2) Why do you think that the Radium girls settled their court case for so little money?

3) The people who were knowledgable of radium's dangers took precautions such as lead screens and masks. Do you think they were concerned about the safety of the young women at their plant? Why do you think they didn't push for implementation of more safety procedures for their fellow workers?

Ethics
Level 1

1) Statistics tell us that 850 out of 1000 lung cancer victims are smokers. If a state has 9600 cases of lung cancer in a particular year, how many of the patients can we predict were smokers.

2) Inferior paint cost $14.95 per gallon, while a quality paint cost $32.50 per gallon. If a gallon will cover 245 square feet, how much money will a contractor save if the inferior paint is used on a wall that is 8 feet high by 345 feet long?

3) If it cost 12.8 million dollars to recall 8 million bottles of medicine, how much would it cost to recall 11 million bottles of medicine?

4) If 2850 people take cable signals without paying for them, how much money will the cable company lose in a year if the monthly cost for cable is $48.75?

5) A carpet company starting selling carpet by the square inch instead of the square yard. Carpet that previously sold for $32 a square yard was now being sold for "only 5 cents per square inch." Which is cheaper, carpet that is $32 per square yard, or carpet that is 5 cents a square inch?

Ethics
Level 2

1) A company that paid its workers $15.50 per hour, moved its factory to a country where they only had to pay the workers $1.35 per hour. By what percent did the companies labor costs drop?

2) A worker who previously was paid $15.50 per hour, had to take a job that paid $5.50 per hour because his employer moved the factory to another country. By what percent did his wages drop?

3) A contractor did $982,000 worth of work in 1997. Because he started using inferior paint, he lost 50% of his business in 1998. He then lost an additional 50% of his business in 1999. How much work did the contractor do in 1999?

4) A cable television company president said that the monthly cost for its 1.5 million paying customers would be $28.50 if nobody was stealing the signals. Because thousands of people are stealing the signals, the monthly cost is$31.35. How many people are illegally taking the cable signal?

5) A check cashing company charges an interest rate of 12% per week. If someone borrows $100, how much will he owe in 10 weeks?

Ethics
Einstein Level

1) A normal life span for a woman is 76 years. If one of the "radium girls" died at 28 years old, by what percent was her life span shortened?

2) In 1970, a certain car cost $1400 to manufacture. If the company spent an extra $11, it would have made the car's gas tank much safer. By what percent would the manufacturing costs have increased if the company fixed the gas tank? (Round to the nearest hundredth of a percent.)

3) Statistics tell us that 850 out of 1000 lung cancer victims are smokers. If a state has 1260 lung cancer patients who are not smokers, how many lung cancer patients would you predict there are in the state who are smokers?

4) A store sold $3,500,000 worth of merchandise in 1992. The store had its sales drop 60% in 1993; 50% in 1994; 40% in 1995; and 30% in 1996. What were its sales in 1996?

5) A store sold "*n dollars*" worth of merchandise in 1992. It had its sales drop 20% in each of the next 5 years. Fill in the total sales in terms of *n* for the five years. The first two are done for you.

1992 --------*n*
1993---------.80*n*
1994---------.80*n* x .80*n* = .64*n*
1995---------?
1996---------?
1997---------?

Chapter 10

Bias is Everywhere

Everyone is susceptible to bias, even the highly educated. Bias distorts the interpretation of scientific experiments, affects the results of surveys, and influences the beliefs that we hold.

Are Some Knee Surgeries Useless?

Almost a quarter million people undergo knee surgery each year to try and gain lost mobility and to alleviate severe pain caused by osteoarthritis. The surgery is typically done by making small incisions in the knee. Once the incisions are made, a tube and camera are inserted into the knee to enable the surgeon to repair any damage that is there.

Even though the patients who had this type of knee surgery appeared to be better off after the surgery, several doctors at the Baylor College of Medicine (Houston) decided to use the scientific method to determine whether the surgery really was useful. Many of the surgeons who performed this kind of knee surgery thought the experiment was a waste of time. Their patients were obviously getting better after the surgery and that was proof enough for them.

The doctors who conducted the experiment divided 180 patients into two groups. Half received the real surgery, while the other half had incisions made in their knees, but no surgery was performed. None of the patients knew whether they had the real surgery or the fake surgery.

Placebo experiments, where patients take fake pills, are very common and are not very controversial. Placebo surgeries are very controversial because they expose the patient to the possibility of infection and other dangerous side-effects. In this experiment, it was decided that the placebo surgeries were worth the risks because if the experimenters showed that knee surgery was no more effective than sham surgery, then thousands of useless surgeries could be avoided each year.

Two weeks after the experiment was completed, the individuals who had the fake surgery were feeling better than the patients who had real surgery. When both groups were checked two years later, they each had an equivalent drop in pain and both groups were walking and climbing stairs better.

This experiment tells us that this type of knee surgery, which is being performed on thousands of people each year, may in fact be useless. The fact that both groups felt better indicates that the placebo effect could have been responsible for the apparent improvements after knee surgery.

The placebo effect is an improvement that happens when someone thinks they are taking a medicine that is supposed to bring about an improvement. It is still too early to tell for sure, but the results of this experiment show that there is a possibility that this type of knee surgery is not as effective as doctors once thought. If other experiments show similar results, then the use of this type of knee surgery will probably decrease dramatically.

The experimenters needed to keep the participants in the experiment "blind" to who actually had real surgery because of bias. If the patients knew they were the ones who had the real surgery, they would have expected to improve.

The beauty of the scientific method is that the truth can be discovered, even if the participants are biased. It is important to realize that everyone is affected by bias -- even doctors and scientists.

Discussion Questions:

1) Explain the meaning of the placebo effect.

2) Why is a placebo such an effective way to relieve pain for some people?

3) The people who were interviewing the patients to see how effective the surgery was did not know who had the real surgery and who had the fake surgery. Why was it important that they not know who had the real surgery.

How Biased Sampling Led to a Wrong Prediction

The outcome of the presidential election of 1936 seemed to be easy to predict. Franklin Roosevelt was a charismatic leader who had helped implement programs that provided jobs for many voters during the Great Depression (New Deal). In addition, he had the opportunity to talk to the American people on a regular basis during his popular radio fireside chats.

Roosevelt's opponent was a Republican named Alfred Landon. Landon tried to argue that Roosevelt's New Deal was not working. He also expressed the opinion that Roosevelt was acting like a dictator and was becoming too powerful. Although Roosevelt was popular, Landon had many supporters throughout the country.

As election day approached, a magazine called the *Literary Digest* sent out 10 million postcard ballots in an effort to predict the winner. After two and a half million ballots were returned, the *Literary Digest* announced with great pride that their prediction was an overwhelming victory for Landon. Their polling showed that Landon would win with 57% of the vote to Roosevelt's 43%.

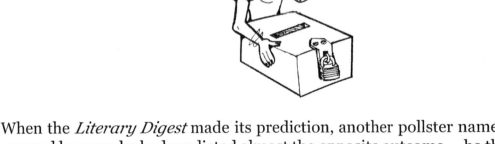

When the *Literary Digest* made its prediction, another pollster named George Gallop was amused because he had predicted almost the opposite outcome -- he thought that Roosevelt would win 56% of the vote. Gallop was very confident that he would be proven correct when the results of the election came in because he knew that there was a serious flaw in the way the *Literary Digest* conducted its survey.

The *Literary Digest* sent "ballots" to 10 million people whose names they found on lists of telephone and automobile owners. Gallop knew that a list of telephone and automobile owners was not an accurate cross-section of the American population. In 1936, only the wealthy and upper middle class owned cars and phones. The *Literary Digest* had ignored much of the lower and middle classes.

Gallop only polled 50,000 people, but he used a representative sample. Because of this, Gallop's polling of a relatively small group of people was much more accurate than the *Literary Digest's* poll of 2.5 million individuals.

The results of the election showed that Roosevelt received 62% of the vote. Gallop became a very popular pollster after this election, while the *Literary Digest* lost its credibility as a polling organization.

I once did a survey to find out if cats are good or bad and my results were that 99.98% of those interviewed thought cats were bad. I think my sample population might have been somewhat biased because I only interviewed mice and rats.

This story shows the importance of using a representative sample when you are taking a survey.

Discussion Questions:

1) The *Literary Digest* had a much larger sample size than the Gallop poll. Why was Gallop's poll more accurate?

2) George Gallop said there was a serious flaw in the way the *Literary Digest* conducted its poll. What was this serious flaw?

3) If you were going to poll people concerning an election in the year 2004, would a telephone book be a good place to find a representative sample of people?

The Discovery that Eliminated Bias

Before 1948, researchers would often get misleading results from their medical experiments because of bias. Sometimes the research consisted of doctors giving their patients a new medicine they thought would help them, followed by the same doctors making judgments about how effective the medicine was. Even when researchers would split a group of patients into a control group and an experimental group, they might consciously or unconsciously place the least healthy patients into the group that wasn't getting the new medicine. This would make it much more likely that the medicine they were testing would appear to be effective.

This bias during the selection process made experiments fairly useless. Any difference between the groups at the end of the experiment would most likely have been due to the health of the patients at the beginning of the experiment.

Another problem researchers had before 1948 was that they usually knew which patients received the experimental treatment, and which ones were in the control group. When the researchers tried to judge how much improvement occurred, they were consciously or unconsciously influenced by their knowledge of who was given the treatment.

You look like you've improved dramatically! My new medicine is a success!

Now, if a doctor claimed that she had a treatment that could make all paralyzed patients walk normally again, she would not have to do a complicated research project to prove that her treatment worked. If paralyzed patients walked out of her office, that would be a pretty strong indication of the effectiveness of her treatment. Unfortunately, the effectiveness of most medical treatments is not so obvious. For example, a heart medication that could prevent 10% of all heart attacks would save thousands of lives, but if a researcher gave it to only a few patients, the lifesaving benefits would not be apparent.

More and more researchers were realizing that bias was a serious problem in the testing of new medicines and treatments. In 1948, British researchers devised a way to study medicines without letting bias ruin their results. In other words, they discovered a new way to find the truth.

Their new method of experimenting was called randomized controlled trials.

That sounds complicated, but if I use the words "randomized controlled trials" when I talk to my friends, they'll think I'm much smarter than I really am.

What randomized controlled trials did was take judgment, hopes, dreams, and expectations out of medical research. It made the testing of medicine a true experimental science.

In the 1948 experiment, the British researchers wanted to know whether streptomycin was an effective treatment for tuberculosis. 107 patients were randomly assigned to two different groups. In this study, 52 people were assigned to a control group and 55 people were assigned to a group that received streptomycin.

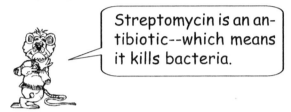

Streptomycin is an antibiotic--which means it kills bacteria.

Not only were the patients randomly assigned to the two groups, but the researchers did not know which patients were getting the medicine and which ones were in the control group. At the end of the experiment, the two groups were examined. One group had 7% of its patients die, while the other group had 27% of its patients die. Radiologists also looked at lung X-rays to see if there were changes in the patients from the beginning of the experiment until the end. What they found was that the group with a 7% death rate had much better lung improvement.

Of course the radiologists also were not told which patients were in the control group and which ones were taking streptomycin.

Only after all the results were in, did the researchers check to see which group was the control and which group was taking streptomycin. The group that was taking streptomycin had the 7% death rate! The world now had a new and scientifically proven treatment for tuberculosis.

Randomized controlled trials helped science take a giant leap forward.

It is interesting that the person who designed the streptomycin study was not a doctor, but a statistician. A mathematician was a hero again!

Discussion Questions:

1) Why is picking patients at random an important part of a good experiment?

2) What is a "randomized controlled study"?

3) Why is it important that the patients not know if they are in the control group or if they are in the group that is getting the medicine?

Bias is Everywhere
Level 1

1) What is a placebo?

2) What is the placebo-effect?

3) Monica wanted to check and see if vitamin C cured colds, so she gave the vitamin to 100 friends when they were sick. A week after she gave each friend the vitamin, she interviewed them to see if vitamin C helped them recover from their colds. Is this good research? Why or why not?

4) Explain how Monica could have used a control group to make her results more reliable?

5) Because Monica is a very strong believer in the powers of vitamin C, she is very worried that her results will be biased. How can Monica "blind" herself to eliminate bias?

Bias is Everywhere
Level 2

1) Lauri conducted a survey to find out how common cheating is at her school. She randomly picked 100 students out of the 1000 who attend her school and interviewed them. She found that none of the 100 students cheated. From this information, Lauri made the claim that there was little or no cheating in her school. Did Lauri do good research? Why or why not?

2) In question #1, how might Lauri have done her research to make her results more accurate?

3) What is a representative sample?

4) A polling company wanted to predict who would win the next presidential election, but it didn't want to go to the trouble of picking 1000 people at random. It decided to do its polling at a union hall after a large meeting. Why is a union hall not a good place to find a representative population to poll? Who do you think they are going to predict will win the election, a Democrat or a Republican?

5) A professor at a university has done several research projects that he says show there is little danger in breathing secondhand smoke. Many people say that the researcher is obviously biased. This researcher does not smoke. Why are some people accusing him of being biased?

Bias is Everywhere
Einstein Level

1) A study concluded that people who visit a doctor take longer to recover from an illness when compared to those who don't visit a doctor. Because of this information, Joel decides that he won't go to the doctor when he has the flu. Explain how Joel's thinking is flawed.

2) A magazine decided to see which hospitals are the safest, so it selected 100 patients from each hospital and followed their progress for 6 months. The magazine was shocked to find out that the hospital with the best reputation for skilled treatment of serious diseases was one of the hospitals with the highest death rate. Does this mean this hospital is one you might want to avoid? Why or why not?

3) Amy and Natalie held opposing views about putting animals to sleep when they are very old and sick. Amy thought that it wasn't right, no matter how sick the animal was. Natalie believed that putting a pet to sleep when it was old and very sick was the kind thing to do. Because they were going to have a debate about the issue in debating club, Natalie and Amy each conducted a survey to see how the public felt about putting pets to sleep. (Euthanasia)

Natalie's survey found that 98% of the public supported her position, while Amy found that 95% of the public supported her position. What is a possible reason for such strange survey results? (They each surveyed 100 people who were picked at random.)

4) What is "unconscious bias"?

5) Marilyn wrote and solved a very tricky probability problem for a national magazine. After the magazine was published, she received hundreds of letters telling her that the answer she gave was wrong. Marilyn's answer was correct, but she was very worried about the country's level of mathematical proficiency because 9 out of 10 people who wrote her said her answer was wrong. Can you use this information to say that it is likely that 9 out of 10 people who solved the probability problem thought Marilyn's answer was wrong?

Chapter 1 : Math and Science Will Tell you the Truth

Level 1:

1) $198 + 12% tax

.06 x $212 = $12.72 tax   Total = $224.72
.12 x $198 = $23.76 tax   Total = $221.76

2) 3 bags

The people weigh 6 x 128 = 768 pounds
The metal floor is 24 square feet, so it weighs 192 pounds.
95 pounds of available weight left

3) 78%

When an average score for three tests is 70%, the total score for the three tests is 3 x 70 = 210. Stan has a total score on two tests of 132, so he needs a test score of 78% on his third test.

4) $6

.12 x $50 = 6

5) 1 in 2

If the question dealt with the probability of a couple with no children having 5 children in a row who are boys, then the answer would be 1 in 32. Each individual birth is 1 in 2.

Level 2:

1) No

The area of the carpet is 30 square yards. (Remember that each square yard is 9 sq. feet.)
270 square feet x .65 pounds = 175.5 pounds

2) No - The tree is 92.57 feet tall

Using Ratios: 36/28 = n/72 Now cross-multiply
28n = 2592   n = 92.57

3) 135 pounds

A cubic yard is 3 ft. x 3 ft. x 3 ft. = 27 cubic feet

4) 234th day

A function machine can be set up to save a lot of time.

13	19	25	3¹	37..1411
day 1	day 2	day 3	day 4	day 5 n

Because there is a space of 6 between each number, multiply each day by 6. (then add 7)

Note: For an explanation on function machines, see page 33 of *Challenge Math.*

$1 \times 6 + 7 = 13$
$2 \times 6 + 7 = 19$
$3 \times 6 + 7 = 25$
$4 \times 6 + 7 = 31$
$5 \times 6 + 7 = 37$

$n \times 6 + 7 = 1411$ $6n + 7 = 1411$ $n = 234$

5) 4 feet

There is 16,000 pounds of available weight.
The volume of the container is 1280 cubic feet.

If the container was filled completely it would weigh $1280 \times 25 = 32,000$ pounds
Because you only want 16,000 pounds, fill it half way, or to the 4 foot mark.

Einstein Level:

1) 1 in 12,271,512 chance

$6/48 \times 5/47 \times 4/46 \times 3/45 \times 2/44 \times 1/43 = 1/12,271,512$ (Cross-reducing is very helpful.)

2) $5710

80,000 miles ÷ 12 mpg = 6667 gallons for the sports car   Cost: $10,000
80,000 miles ÷ 28 mpg = 2857 gallons for the practical car   Cost: $4286

3) 1 in 649,740

Your first pick has a chance of 20/52 because there are 20 cards in the deck that will start you on your way to a royal flush. After your first pick, there are only four good cards left in the deck.

$20/52 \times 4/51 \times 3/50 \times 2/49 \times 1/48 = 1/649,740$ (Cross-reducing is helpful.)

4) $3145

1st year: 1.229 x 400 = $491.60
2nd year: $491.60 x 1.229 = $604.18
3rd year: $604.18 x 1.229 = $742.53
And so on

5) $2,048,000

The money doubles every 6 years

0 years: 1000
6 years: 2000
12 years: 4000
And so on

Chapter 2 : Counterintuitive

Level 1:

1) 50 cents

Ball: n
Glove: n + 9
Equation: 2n + 9 = 10   n = $.50

2) 10^{86}

3) It is the same pressure either way.

4) Because it looks like the sun is going around the Earth.

5) They will hit at the same time.

Level 2:

1) Both are equally violent.

The severity of a crash depends on how quickly the person inside goes from 60 miles per hour to 0 miles per hour. In both situations, the deceleration is almost identical.

2) 4 mph

Most people will write 4.5 mph, but this is wrong. Because Phil spent more time traveling at 3 mph, this speed must be counted more. In this problem, he spent twice as much time going 3 mph as 6 mph so it would be counted twice as much. There are three blocks of time as shown below.
3 + 3 + 6 = 12   12 ÷ 3 = 4

3) It is impossible.

Using the formula: *Distance = Speed x Time*
The distance will be 2 miles and the speed is supposed to average 60 mph.
2 = 60 x n   n = 1/2 hour
The amount of time will be 30 minutes, but the driver has already used that amount of time.

4) Choice 2: A raise of $400 each six months

Look at the following chart that breaks the payments down into 6 month sections:

	Option of $400 each 6 months	*Option $1000 each year*
1st year: first 6 months	$10,000	$10,000
next 6 months	$10,400 ($400 raise)	$10,000
2nd year: first 6 months	$10,800	$10,500
next 6 months	$11,200	$10,500
3rd year: first 6 months	$11,600	$11,000
next 6 months	$12,000	$11,000
4th year: first 6 months	$12,400	$11,500
next 6 months	$12,800	$11,500

As you can see, Mark would be much better off taking the $400 raise every 6 months. The reason the brain has a hard time with this problem is that it jumps to the conclusion that a $400 raise every 6 months in the same as an $800 per year raise--it is not!

5) No- His chance of being kicked out of medical school remains at 1 in 3. Stacie's chance of being expelled, however, is now 2 in 3.

Stacy and Laura as a group had a 2 in 3 chance of one of them being expelled. Chris had a 1 in 3 chance. The dean always would pick the girl from the group of two who was graduating. The group still has a 2 in 3 chance of having the expelled student in it, and because Stacie is in that group, she has a 2 in 3 chance of being the student who is expelled.

(If you don't understand, don't feel bad. Some of the smartest people in the world get this one wrong.)

Einstein Level:

1) 2 in 3

This is a very counterintuitive problem. The best way to think of this is to ask yourself, before you pick a coin, what is the probability of picking a coin that has the same thing on both sides of the coin. (A heads or tails showing is equally likely so that doesn't change the original probability.) The answer of course is 2 in 3. Now you have a coin in your hand with heads showing. What is the possibility of it being a double sided coin. 2 in 3!!

2) Let's say Player B shoots one for 10 out of his first 10 shots. Now they are both shooting 10 for 40 or 25%. All the problem does is have Player B take more shots. The rest of his 20 shots he is 5 for 20, which maintains his 25% average.

3) No advantage to switch

At first glance, it appears that you should switch. After all, there may be $450 or $1800 in the other box. It seems that you have an equal chance of losing $450 or gaining an additional $900. But that line of thinking would tell you to switch every time, which means that when you are picking Box A, you know you will end up with Box B. If you pick Box B, you know you will end up with Box A. Either way, you are just randomly picking a box.

4) 1 in 3

Let's say that the boys and girls are coins. I ask you to throw two coins into the air and let them land on the table. I ask you one question. Is at least one of the coins heads? If you say yes, then I know the arrangement of coins on the table is either:

1) Heads   Tails
2) Tails   Heads
3) Heads   Heads

If at least one of the coins is heads, the chance of the other coin being heads is clearly 1 in 3. If one of the children is a boy, the chance of the other also being a boy is 1 in 3.

5) About a 50 -50 chance.

Remember, the question was not talking about a certain date. It was asking if any of the children would match with any of the others.

Chapter 3 : Occam's Razor

Level 1:

1) The most likely explanation for the sound of hooves is horses, not something exotic like zebras.

2) No, it was a pony that was covered with a buffalo skin.

3) It was a popular exhibit until an unexpected rain occurred and washed the white paint off of the Light of Asia.

4) An egress is an exit. Once visitors to the museum went through the door, they had to pay another admission to get in.

5) The picture could have been taken at the top of his hop.

Level 2:

1) The willingness to believe, even if the evidence is weak or nonexistent.

2) The invasion was a Halloween show hoax by Orson Wells called *The War of the Worlds.* At the end of the terrifying report on the invasion, the deep voice of Orson Wells announced: "You have been listening to Orson Wells and the Mercury Theater of the Air in a radio adaptation of *The War of the Worlds.*"

3) Blackburn had distracted the scientists and slipped the messages to Smith on pieces of cigarette paper. Smith was able to read the messages with a luminous stone that he kept in his pocket. The two men engaged in this deception because they wanted to demonstrate the fact that scientists who really wanted to believe something are easily fooled.

4) Parsimony means excessively frugal or stingy. The rule of Parsimony relates to being frugal with explanations---making only simple explanations.

5) The horoscope fits almost everyone. Almost all people feel they are misunderstood or that they are a good person at heart. In addition, very few people like to have others mad at them.

Einstein Level:

1) Richard Feynman thought that the most down to earth explanation for UFO's was that they were the result of people playing jokes or engaging in trickery. He did not think that "visitors from outer space" was a simple explanation.

2) After Dr. Adams died, several of his machines were opened and found to contain wires, condensers, and electrical parts---all disconnected and arranged in a meaningless fashion. Dr. Adam's career was based on a fraud that affected hundreds of doctors and thousands of patients.

3) It is very easy to "mentally" bend a spoon. Before a performance, the spoon is bent back and forth until it is just about to break. Now any slight movement will cause the top to lean and then break off.

4) While most people who claim to have a crying statue don't allow an inspection of the statue, it is very easy to explain the tears. If you fill a glazed ceramic statue with water and then scratch the glazing away from the eyes, water will seep through the statue at that point.

5) Bacteria in water can come and go. The bacteria level probably would have dropped whether the pyramid was there or not.

Chapter 4 : Mistakes and Frustration

Level 1:

1) 144 square inches

2) Mass: How much stuff something is made of. Mass doesn't change when the force of gravity changes.

Weight: Measurement of the force of gravity. Weight changes when gravity changes.

3) 16

4) $100,000

The median is the middle number when the list is put in order from the smallest to the largest.

5) Speed is the measurement of how fast an object is moving.

Acceleration measures the change in speed.

6) 500.02

7) A foot is a measurement of length. A square foot is a measurement of area. A cubic foot is a measurement of volume.

8) .0100

9) .001 kilometers in a meter.   (1/1000 of a kilometer in a meter)

10) You can't tell.

If n is a negative number, then $2n$ will be smaller than n.

Level 2:

1) 1728 cubic inches

12 x 12 x 12 = 1728

2) Mass refers to the amount of "stuff," while volume refers to the size.

3) 3.14

The ratio of the circumference to the diameter of any circle is pi.

4) 2

.5% is the same as 1/2%   1/2% of 400 is 2

5) 10

6) No, 1 is not a prime number.

7) No, if n is negative, then -n will be positive

8) -273°C

Temperature is a measurement of the speed with which molecules move. When motion stops, the temperature cannot go any lower. All motion stops at absolute zero, which is -273 C.

9) Bacteria are bigger than viruses and can be killed with antibiotics. Viruses are very small and antibiotics are not effective against them.

10) 312.5 pieces

If you had an answer of 200, that is obviously not correct because each piece is less than a pound and the chocolate is 250 pounds. Sometimes the brain can't think properly when there are fractions in problems, so you can pretend the chocolate weighs 10 pounds and then ask yourself: "How many two pound pieces can be cut from 10 pounds"? The answer is obviously found by division, so go back to the original problem and divide.

Einstein Level (1)

1) Newton (Most people say kilogram or gram, but that is not technically correct)

2) Slug

3) $37.50

50% off of $100 is $50	New price is $50
25% off of the $50 is $12.50	The new price is $37.50

4) $848

1.07 x total sales = 856   So the total sales equal 856 ÷ 1.07 or $800
1.06 x $800 = $848

5) $5^1 = 5$ $5^0 = 1$ (All numbers to the zero power are 1)

6) Natalie did the prime factorization of 60, she didn't find the prime factors. They are 2,3,5

7) $.76

$$\frac{1.32\ Canadian}{1\ American} = \frac{1\ Canadian}{n\ American}$$ $1.32n = 1$ $n = .76$

8) One hour and 20 minutes

Think how much of the fence can they each do in an hour? Bill can do one half while Steve does one fourth. Together they do 3/4 of the fence in an hour. Each quarter takes them 20 minutes, so the whole fence can be done in 1 hour and 20 minutes.

9) -40 degrees

10) 9/32

Change the fractions into 32nds. Halfway between 6/32 and 12/32 is 9/32.

Einstein Level (2)

1) 215,056 legs (Don't forget the farmer's legs.)

2) 101.5%

If there are six tests and Ted wants to average 86%, then the six scores must add up to 6 x 86 = 516. The tests taken so far add up to 414.5.

3) 576 square inches

Call the width of the figure *n*. The length would then be 4*n*.   *4n + 4n + n + n* = 60 inches

n would then equal 6. The square is 24 inches by 24 inches or 576 square inches.

4) 5 1/3 days

Think one student. The food would last one student 96 days. (12 x 8) The food would last 18 students 96 ÷ 18, or 5 1/3 days

5) Thursday

6) 385 squares

Pattern------------- Big squares: 1
Next biggest size: 4
Next biggest size: 9
Pattern is perfect squares: 1 + 4 + 9 + 16 + 25 + 36 + 49 + 64 + 81 + 100 = 385

7) 20%

150 faces are painted. 125 cubes have 750 total faces.

8) 8

Look for a pattern: 8 to the 1st power ends in 8
8 to the 2nd power ends in 4
8 to the 3rd power ends in 2
8 to the 4th power ends in 6
8 to the 5th power ends in 8   and so forth

9) 4:24

LCM of 8 and 9 is 72. They will meet in 72 minutes.

Einstein (3)

1) 46 minutes 40 seconds

$$\frac{7.2 \ miles}{42 \ minutes} = \frac{8 \ miles}{n} \qquad 7.2n = 336$$

2) 203

$3n + 11n + 29n = 301$
$43n = 301$
$n = 7$
$7 \times 29 = 203$

3) 12 inches

4) 79 cents

5) 120

Least Common Multiple of 6, 8, and 10 is 120

6) 64

7) $157.50

Volume is 9 yards x 3 yards x 1/6 yards = 4.5 cubic yards.

8) 125

Each number is a perfect cube.

9) 6.1 inches

The volume of the original cake is 390 cubic inches.

The pan with a 9 inch diameter has an area of 63.585 square inches. To get a volume of 390 cubic inches, you must multiply 63.585 by 6.1

Einstein (4)

1) 1000%

.0005 ÷ .00005 = 10   10 = 1000%

2) 96

$$\frac{1}{8}n + 50 = \frac{1}{4}n + 38$$

3) 12:30

The broken clock goes 48 minutes for each 60 of the good clock: $\dfrac{Broken\ 48}{Good\ 60} = \dfrac{Broken\ 120}{Good\ n}$

n = 150 minutes for the good clock.

4) 137 miles

D = R x T   D = 36 x 3.8 hours

5) 5

Change all to 12ths

$$Average\ of\ \frac{6}{12} + \frac{4}{12} + \frac{3}{12} + \frac{2}{12} + \frac{n}{12} = \frac{4}{12}$$
$$(6+4+3+2+n) \div 5 = 4$$
$$(15+n) \div 5 = 4\ \ so\ \ n\ \ must\ \ equal\ \ 5$$

6) 19 days

4560 miles equals 24,076,800 feet   Divide by 880 = 27360 minutes=456 hours=19 days

7) 5,865,696,000,000 miles

31,536,000 seconds in a year multiplied by 186,000 miles equals 5,865,696,000,000 miles

8) 11

1000 x 55% x 40% x 10% x 50% = 1000 x .55 x .4 x .1 x .5 = 11

9) 1/7

Einstein (5)

1) 11.3 miles per hour

Because Jill spends more time traveling at lower speeds, there must be more blocks of time at the slow speeds.
(4 mph has 12 blocks because it is 12 times slower than 48.)
(16 mph has 3 blocks because it is 3 times slower than 48 mph.)
(48 is counted twice because Jill goes twice the distance at that speed.)

(4 x 12) + (16 x 3) + 48 + 48 = 192   192 divided by 17 blocks = 11.3

2) 1/36
$$\frac{1}{1}x\frac{1}{6}x\frac{1}{6}=\frac{1}{36}$$
(Isaac's first roll will always be part of 3 of a kind.))

3) 2:15
$$\frac{56\ minutes\ (broken)}{60\ minutes\ (Good)}=\frac{126\ minutes\ (broken)}{n}$$ n (good clock) went 135 minutes

4) $7\frac{17}{44}$

5) 21 dimes
Number of nickels: n Value of nickels: $5n$
Number of dimes: $7n$ Value of dimes: $70n$
number of quarters: $21n$ Value of quarters: 525n

6) 19 cubic inches

The volume of a cylinder that is 11 inches high with a circumference of 8.5 inches is 63.28 cubic inches. The volume of a cylinder that is 8.5 inches high with a circumference of 11 inches is 81.89 cubic inches.

7) 43,046,721
These are all numbers raised to the forth power. 81 to the forth power is 43,046,721

8) .15%

5% of 20% of 15% equals   .05 x .2 x .15 = .0015 which is equal to .15%

9) .000,000,000337
When you divide a number by 1,000,000, you need to move the decimal place six places to the left.

Chapter 5 : It is Important to Keep an Open Mind

Level 1:

1) Train

2) Hot-air balloon

3) Meteorites

4) Personal computers

5) Typewriter

Level 2:

1) The circulation of blood

2) Vaccinations

3) Galileo

4) Radio communication

5) Telephone

Einstein Level:

1) Galvani accidently discovered that when an electrical current touches a frog's leg, it forces the leg to twitch.

2) Ice-making machine

3) Steam-engine boat

4) The theory of continental drift. Wegener thought that the continents were joined millions of years ago.

5) Dr. Joseph Lister

Chapter 6: Healthy Skepticism

Level 1:

1) N-rays were a real "discovery", but they only existed because Blondlot and many other scientists thought they did. N-rays were a classic example of bias and "seeing what one expects to see." N-rays were and are totally imaginary.

2) A bird was used. Augury is obviously not legitimate and fortunately it is no longer done.

3) Yes. The common name for a "radio detection and ranging" system is radar.

4) Simple scientific tests have shown that these pyramids have no special powers.

5) A 4-week study at the Mayo Clinic was conducted in 2002. Some patients wore the special bracelet and others wore a placebo bracelet. The improvement in pain scores was identical for both groups, which showed that the bracelet did not have any special powers.

Level 2:

1) Even though thousands of people strongly believe in reflexology, there is no scientific evidence to show that it is legitimate.

2) No one can levitate, but many people can make the illusion so powerful that they really appear as though they have that ability.

3) No scientific evidence supports the idea of phrenology. The theory was proposed in 1796 and is still practiced by some people today.

4) They are telling the truth. The device they are talking about is a simple antenna that goes on top of a television. (Sometimes called rabbit ears) They were very common before cable was introduced, and are still used today.

5) Radionics and various other forms of wave-therapy have been around for hundreds of years. Scientific testing has shown that radionics and the theories behind it are quackery. Proponents of radionics are usually well-intentioned individuals, but some are involved in deliberate fraud.

Einstein Level:

1) The available scientific evidence would say that the proponents of "oxygenated water" are either involved in well-intentioned foolishness or outright fraud. The oxygen content of the air has not changed and remains at about 21%. In addition, the body extracts the oxygen that it needs from the air and returns surplus oxygen to the lungs to be exhaled. (Even if these pills did contain extra oxygen, humans cannot extract any meaningful amount of oxygen from the stomach) The Federal Trade Commission has filed several suits against companies that market "oxygenated water." In its long list of complaints, the FTC said that these products appear to be nothing more than salt water.

2) There is an extraordinary amount of scientific research to back the claims made about acetylsalicylic acid. Another name for acetylsalicylic acid is **aspirin.**

3) Iridology is a real belief system, but it is quackery. It has been scientifically tested several times and has always failed. Iridology should not be confused with the proven medical practice of looking into the eye for symptoms of certain diseases.

4) Scientific studies on crystals show that they are beautiful rocks and nothing more. They do not have psychic powers or have the ability to heal.

5) Many people claim to have invented perpetual motion machines. Unfortunately, our understanding of physics tells us that perpetual motion machines are impossible. When someone claims to have a perpetual motion machine, he has either made an honest mistake or he is involved in fraud.

Chapter 7 : Don't be Fooled by Statistics

Level 1:

1) Knowing the average salary is not enough information. Company B may have one person who is paid $100,000 while the rest of the employees are paid $20,000.

2) What was the population of the city that had 32 murders in 5 years?

3)

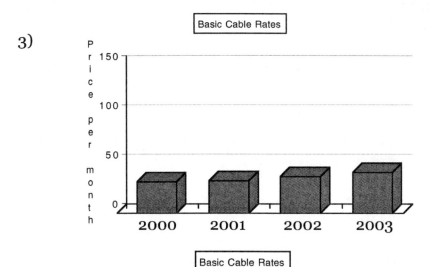

4)

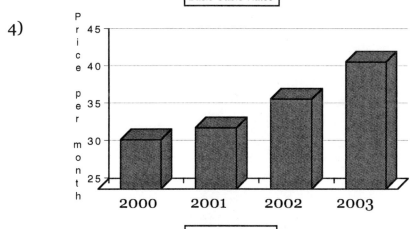

5)

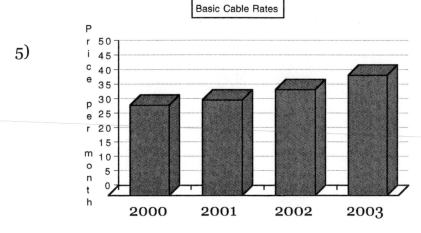

Level 2:

1) A 50% raise when the salary is $25,000 is only $12,500. The 50% decrease was $25,000.

2) The graph stops at 60%. This gives a visual impression that achievement is near the top.

3)

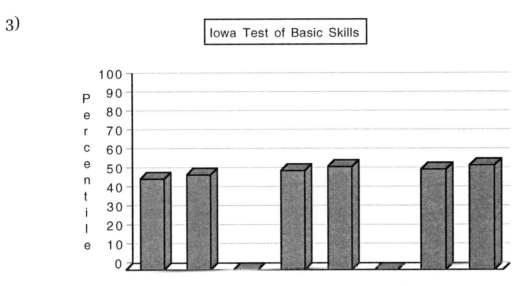

4) No, the scale starts at 20,000 to exaggerate the differences.

5) Painting bull's-eyes around arrows. Seth looked at the crime statistics and picked the ones that went down. He then took credit for the drop. Seth doesn't mention what happened in the many other categories of crime.

Einstein Level:

1) No - When the radius of a circle doubles, the area of the circle quadruples.

2) No - The second triangle will have four times the area of the first.

3) No - When the radius of a sphere doubles, there is an eightfold increase in volume.

4) No - The local store could have looked through all the prices and picked out the few where they were cheaper. They were probably "painting bull's-eyes around arrows."

5) Janelle's percentile score tells her that she did better than 90% of students her age. Percentile scores do not measure an "amount" of intelligence. A person who is taller than 90% of people his age is not twice as tall as someone who is taller than 45% of the people who are his age.

Chapter 8 : Cause and Correlation

Level 1:

1) The rooster crowing and the sun appearing occur at the same time, but the rooster's crowing doesn't cause the sun to rise.

2) B If it never rained, we don't know how effective his invention is.

3) C The researcher must check to make sure that most people don't have these old brain injuries. His theory only makes sense if old brain injuries are unique to violent people.

4) The reason more fatal accidents occur within 30 miles of one's home is because almost all driving is done within 30 miles of one's home.

5) False

Level 2:

1) You cannot make this kind of judgement with the information given. In this case, Country A is so poor that the average life-span is 45 years. Because cancer is usually a disease of old age, a population that dies young will have very little cancer.

2) The two clusters are probably a hospital and a nursing home.

3) Many people retire to Florida and Arizona. If there are more elderly people in those two states, they are more likely to have a higher death rate.

4) No. In all likelihood, there are many more Ketter bikes sold than Sloan bikes, so the accident rate would obviously be higher. If Sloan advertised that their "deaths per 1000 bikes sold" was better than Ketters, then the information would be meaningful.

5) A If the sweetener causes a strong craving for carbohydrates, then it is unlikely that it will lead to weight loss.

Einstein Level:

1) Vaccinations are typically given to children around the same age that the symptoms of autism appear. Because of this correlation, it is very easy to jump to the conclusion that the vaccinations are the cause of the autism.

2) The people who had brain tumors probably drank milk in addition to using cell phones. That doesn't mean the milk caused their brain tumors. You need to know the rate of brain cancer in the population as a whole and then compare it to the rate of brain cancer in cell phone users.

3) Soldiers, for the most part, are young and not likely to die. New York City has a large mix of people including many who are elderly.

4) No. Country A may be a very wealthy country while Country B may have only rock and dirt houses that not only don't burn, but aren't worth much if they do.

5) No. The school board is confusing cause and correlation. Just because the teachers are in a school with low achievement does not mean they are the reason for the low achievement. An overwhelming amount of research points to social and economic conditions as the most important determining factors in school achievement.

Chapter 9 : Ethical Decision Making

Level 1:

1) 8160

850 out of 1000 is 85%   85% of 9600 is 8160

2) $210.60

An 8 foot by 345 foot wall is 2760 square feet.
2760 ÷ 245 = 11.3 gallons (12 gallons)
$14.95 x 12 = $179.40 $32.50 x 12 = $390 $390 - $179.40 = $210.60

3) 17.6 million dollars

$$\frac{12.8 \ million \ dollars}{8 \ million \ bottles} = \frac{n}{11 \ million \ bottles}$$ $8n = 140.8$ $n = 17.6$ million dollars

4) $1,667,250

$48.75 x 12 x 2850 = 1,667,250

5) Carpet that is $32 per square yard

There are 1296 square inches in a square yard (36 x 36). Carpet that is $32.00 per square yard would cost 3200÷1296 or 2.47 cents per square inch.

Level 2:

1) 91%

Remember, the percent of increase or decrease is the amount of change divided by the original.

The drop was $14.15   The original was $15.50   14.15 ÷ 15.50 = .9129

2) 64.5%

The drop was $10. The original was $15.50.   10 ÷ 15.50 = .645

3) $245,500

50% of 982,000 = $491,000   50% of 491,000 = 245,500

4) 150,000 people

$31.35 x 1.5 million is the amount of money taken in by the cable company. ($47,025,000)

$28.50 times what number would still bring in $47,025,000?

$47,025,000 ÷ 28.50 = 1,650,000

150,000 people are taking the signal

5) $310.58

1 week: 1.12 x 100 = $112
2 weeks: 1.12 x 112 = $125.44
3 weeks: 1.12 x 125.44 = $140.49
4 weeks: 1.12 x 140.49 = $157.35
5 weeks: 1.12 x 157.35 = $176.23
6 weeks: 1.12 x 176.23 = $197.38
7 weeks: 1.12 x 197.38 = $221.07
8 weeks: 1.12 x 221.07 = $247.60
9 weeks: 1.12 x 247.60 = $277.31
10 weeks: 1.12 x 277.31 = $310.58

Einstein Level

1) 63%

48 years shorter   48 is what part of 76? $48/76 = 48 \div 76 = 63\%$

2) .79%   (Less than 1%)

Percent of increase is the amount of rise divided by the original $11 \div 1400 = .0078571$

3) 7140

If 85% of lung cancer victims are smokers, then 15% are not smokers.

15% of n = 1260   (Where n stands for the total number of lung cancer patients.)

$.15n = 1260$ $n = 8400$ 85% of 8400 = 7140

4) $294,000

Remember that if sales drop 60%, they are 40% of the level they started at.

(1993) 40% of 3,500,00 = 1,400,000   (1994) 50% of 1,400,000 = 700,000

(1995) 60% of 700,000 = 420,000   (1996) 70% of 420,000 = $294,000

5) 1995= $.512n$ 1996 = $.4096n$ 1997 = $.32768n$

If the sales drop 20% each year, then each year's sales are 80% of the previous year.

1995 = .8 x $.64n$ or $.512n$

1996 = .8 x $.512n$ or $.4096n$

1997 = .8 x $.4096n$ or $.32768n$

Chapter 10 : Bias is Everywhere

Level 1:

1) Fake treatment

2) The placebo effect is when a person improves because she thinks she is receiving a treatment or medicine.

3) No. The friends are too likely to experience the placebo effect. In addition, they may tell Monica what they think she wants to hear.

4) She could have given 50 people vitamin C and the other 50 people a sugar pill. Neither group should know whether they are getting the vitamin C or the sugar pill.

5) Monica should not know who is getting the real vitamins and who is getting the placebo until the experiment has been completed.

Level 2:

1) The sample size was fine, but Lauri couldn't really be sure the students she interviewed told her the truth.

2) She could have sent the questionnaire anonymously or asked questions such as: Have you ever seen or heard of fellow students cheating?

3) A representative sample is a cross-section of the population.

4) Union members are usually Democrats.

5) The money that he receives for his research is from tobacco companies.

Einstein Level:

1) The reason people who visit doctors take longer to recover is almost certainly because those who are more ill are the ones who feel it is necessary to get medical attention.

2) This hospital, because it is known for its skill in treating serious diseases, is the one people go to when their local hospital can no longer help them. It is not surprising that this hospital has a high death rate----it treats the most seriously ill!

3) Although there could be many different reasons for the surveys to show such different results, the most likely reason was that their questions encouraged biased responses.

Natalie's survey question could have been: Do you think people should have the option of putting old, sick, suffering animals to sleep?

Amy's survey question could have been: Do you think that it is right to kill pets when you don't want them anymore?

4) Unconscious bias is a bias that an individual is not aware of.

5) No. The people who wrote Marilyn were not a representative sample. It is unlikely that someone would write to Marilyn to say he or she agreed with her solution. The people who were most likely to write were those who strongly disagreed with Marilyn's answer----a biased sample!

Bibliography

Gonick, Larry and Art Huffman. (1991). *The Cartoon Guide to Physics.* New York: HarperPerennial.

Gonick, Larry and Woollcott Smith. (1993). *The Cartoon Guide to Statistics.* New York: HarperPerennial.

Huber, Peter W. (1991). *Galileo's Revenge.* New York: Basic Books.

Lovell, Jim and Jeffrey Kluger. (1994). *Lost Moon.* New York: Houghton Mifflin Company.

Park, Robert. (2000). *Voodoo Science.* New York: Oxford University.

Petroski, Henry. (1992). *To Engineer is Human.* New York: Vintage Books.

Randi, James. (1995). *An Encyclopedia of Claims, Frauds, and Hoaxes of the Occult and Supernatural.* New York: St. Martin's Griffen.

Savant, Marilyn Vos. (1996). *The Power of Logical Thinking.* New York: St. Martin's Press.

Shermer, Michael. (1997). *Why People Believe Weird Things.* New York: W.H. Freeman and Company.

Tabori, Paul. (1993). *History of Stupidity.* New York: Barnes & Nobles Books.